William E. Shuckard, W. Spry

The British Coleoptera Delineated

Consisting of Figures of all the Genera of British Beetles

William E. Shuckard, W. Spry

The British Coleoptera Delineated
Consisting of Figures of all the Genera of British Beetles

ISBN/EAN: 9783337144067

Printed in Europe, USA, Canada, Australia, Japan

Cover: Foto ©berggeist007 / pixelio.de

More available books at **www.hansebooks.com**

THE

BRITISH COLEOPTERA

DELINEATED,

CONSISTING OF

FIGURES

OF

ALL THE GENERA OF BRITISH BEETLES,

DRAWN IN OUTLINE BY

W. SPRY, M.E.S.

EDITED BY

W. E. SHUCKARD, Libr^N. R.S.

AUTHOR OF

Essay on the Fossorial Hymenoptera," and "The Elements of British Entomology."

LONDON:

HENRY G. BOHN, YORK STREET, COVENT GARDEN.

1861.

PREFACE.

—◆—

In presenting this work to Entomologists, I do it with the fullest confidence of its being well worthy their acceptance. The deficiency of a similar undertaking has long been felt, and this can be fully appreciated when we reflect how very inadequate descriptive language is to convey a clear and distinct notion of the ever-varying modifications of form observable in the insect world. Considering its scope and object, this work, as one of illustrations in this science, and containing 638 figures, is certainly the most complete that has yet issued from the press; and in artistical execution I will affirm—and which I, sharing in no portion of that merit, may do without egotism—that there is none yet published that surpasses it; and for delicacy of outline it is unrivalled. It is very superior in execution to Panzer and Olivier, whose plans were too vast to admit of completion : and although it wants colour to vie with the best, yet in careful accuracy I am convinced it is not by that even surpassed. It has the advantage of being done throughout from the originals, which, as many of the genera figured are unique, could of course only be accomplished by the kindness of the several Entomologists in whose possession those unique and rare insects existed; or who, with genuine scientific liberality, furnished me for my own collection with the means of supplying deficient forms. To Messrs. Bennet, Desvignes, Hope, Little, Matthews, Newman, Rudd, Stephens, Walton and Waterhouse, mine and the work's best thanks are due for the unlimited confidence with which they entrusted out of their own possession their several rarities, and which, I am happy to say, met in no instance with any casualty. To Mr. Curtis I am also indebted for my own means of examining and for Mr. Spry's liberty of delineating at his residence a form extant in no other British Collection ; and likewise to the Officers of the Zoological Department of the British Museum for the urbanity with which they met and forwarded my objects in the several instances I had occasion to examine their rich stores. Having thus paid the debt of gratitude due for being enabled to complete within the short period of twelve months so extensive an undertaking, for which the praise must be bestowed upon the incessant and unflinching perseverance of Mr. Spry, it is next requisite to give some account of my own peculiar portion of the task, as editor.

The primary object was to produce a work that should be cheap as well as useful, and to compass the former it was not possible to introduce dissections of the parts of the mouth, which I freely admit are in many instances requisite to show the distinctive differences of certain genera where forms are closely allied, or where in long genera the species range divergently from the types. This would of course have incurred a great cost of time, the value of which must necessarily have been thrown upon the work, and which thus, although it would have acquired to a certain extent a greater degree of utility, would from the additional expense have limited considerably its circulation, for the Enomologists to whom a work of forms is most acceptable and useful are the young. A reciprocating task is therefore left to be performed between description and figure : to fulfil the former several English works already exist, and it would have intrenched considerably upon the property of those works, (the entomological public being so small,) had the letter-press of the present extended to a full generic description, as I consider a synoptical description worse than useless. I therefore chose the only alternative of giving nothing more than an outline of my own views of the distribution of British Coleoptera, and a very brief specific description to exhibit the colouring of the species figured. And even had the generic descriptions been added and made full, it could not have precluded the necessity of occasionally referring to other works for information which it would not have contained without a seriously detrimental increase of bulk and cost ; and I have, therefore, preferred making it supplemental to all works upon the subject, to which it will be found a desirable, important and even necessary auxiliary.

I have added a selection only of the foreign genera that have occurred in Britain: indeed, it was of but little moment if any of these were figured; but I did so, thinking it might be acceptable to many, although there can be scarcely a doubt that they have all been accidentally introduced. In the census of species I have exercised the discretion I thought requisite, either in limiting or interrogating the repletion with which our lists are swollen. Truth should be the object of all inquiries, and it is quite as prejudicial to the advance of science to exaggerate as it is to depreciate the extent of our natural productions.

W. E. SHUCKARD.

CHELSEA,
July, 1840.

CORRIGENDA.

Page 6.—Genus [44] and [50]. Transpose the names, both generic and specific, as [44] should be *Olisthopus*; and alter at Pl. 6, fig. 3, the name *Odontonyx* to *Olisthopus*. Subsequently to *Olisthopus* being figured I discovered that no insect had yet been found combining the characters of *Olisthopus*, in form and trophi, with the denticulated tarsi of *Taphria* (*Synuchus*). This Mr. Stephens has since admitted to me to have been an accidental error. I have therefore in Pl. 7 necessarily omitted *Odontonyx*, as not extant in nature, but have inserted its name to show that it is purposely left out.

Page 28.—Alter the numbering of fig. 6 into 8, 7 into 6, and 8 into 7.

Page 49.—Alter the reference of Genus [421], *Tenebrio*, to fig. 3, instead of fig. 2; and the reference of the next genus [422], *Stene*, to fig. 4 instead of 3.

> Note that *Stene* is identical with *Tribolium* figured in the Supplement, Pl. 2, fig. 5, which from haste I did not discover until too late to cancel it. *Tribolium* is the name therefore to be adopted, as it was described by Mr. Macleay in the "Annulosa Javanica" in 1825; and although not originally a native, as it now breeds here, like other domestic insects that have been introduced with merchandise, it may perhaps be admitted in our lists.
>
> The genus *Ecanus* I have been obliged to omit, as I could not obtain either the unique British or even a foreign specimen of it.
>
> The publication of Dr. Erichson's works since the commencement of the present have introduced several new genera amongst the Staphylinidæ, by the dismemberment of the old genera, *Anthobium* and *Lesteva*—the differences however lie chiefly in the trophi. I have therefore thought it not requisite to give figures of them, as I could not introduce them into their proper places.

BRITISH COLEOPTERA
DELINEATED.

COLEOPTERA. Linnæus.

Section I. PENTAMERA. Latreille.

Subdivision I. ADEPHAGA. Clairville.

Tribe I. GODEPHAGA. Macleay.

Race I. EUPTERINA. Kirby.

Family [1]. Cicindelidæ. Leach.

Genus [1]. Cicindela. *Lin.* Three first joints of anterior tarsi dilated in the male. Seven species.

C. campestris, *Lin.* Above of a rich silky green; labrum white; margins of the head and thorax, side of the breast, legs, and basal joints of antennæ of a copper red; five white spots on the external margins of each elytron, one of them humeral, a white spot on disk opposite the third lateral one, and in addition in the female opposite the second lateral one, a minute black spot nearer the suture; five lines; sandy sunny situations. *Pl.* 1, *fig.* 1.

Race II. EUTRECHINA. Kirby.

Family [2]. Brachinidæ. Macleay.

Genus [2]. Drypta. *Fab.* Three first joints of anterior tarsi dilated in the male. Six species.

D. emarginata, *Oliv.* Green inclining to a bright blue; antennæ (except the tip of their first joint) mandibles and legs red; head and thorax coarsely punctured, four lines; under stones. Southern coasts, rare. *Pl.* 1, *fig.* 2.

B

Genus [3]. POLISTICHUS. *Bonelli.* One species.

P. fasciolatus, *Rossi.* Pitchy brown, a short central stripe on each of the elytra, as well as the antennæ, legs and abdomen reddish ; four and a half lines ; under stones, eastern coasts. *Pl.* 1, *fig.* 3.

Genus [4]. CYMINDIS. *Latr.* Eight species.

C. angularis, *Gyll.* Pitchy : thorax ferruginous ; humeral angles, and external margin of elytra, and the legs testaceous ; four lines ; under stones, Dorking. *Pl.* 1, *fig.* 4.

Genus [5]. BRACHINUS, *Web.* Five species.

B. crepitans, *Lin.* Ferruginous red ; mouth and the third and fourth joints of antennæ pitchy ; elytra black with a bluish or greenish tinge and subcostate ; four lines ; under stones. *Pl.* 1, *fig.* 5.

Genus [6]. ODACANTHA. *Payk.* One species.

O. melanura, *Lin.* Greenish blue ; base of antennæ and palpi ; legs (except their knees and tarsi which are black) and elytra testaceous, a bluish irregular patch at the apex of the latter ; three lines ; reeds, margins of marshes, Cambridge. *Pl.* 1, *fig.* 6.

Genus [7]. DEMETRIAS. *Bonelli.* Differs from Dromius by the penultimate joint of the tarsi being bilobate. Four species.

D. imperialis, *Meg.* Pale yellow-testaceous ; head and knees black ; elytra with a sutural pitchy mark dilated at the base and hastate at the apex, and another marginal one, on each side towards the apex ; three lines ; fens, Cambridge. *Pl.* 1, *fig.* 7.

Genus [8]. DROMIUS. *Bonelli.* Differs from Demetrias by the penultimate joint of the tarsi being simple. Twenty two species.

D. agilis, *Fab.* Reddish-pitchy ; thorax and head rather the lightest ; antennæ and legs testaceous ; three lines ; in moss and under bark. *Pl.* 1, *fig.* 8.

Genus [9]. LEBIA. *Latr.* Three species.

L. turcica, *Fab.* Black ; the mouth, antennæ and thorax red ; the elytra with a pale humeral mark ; legs testaceous ; two and a half lines ; very rare. *Pl.* 2, *fig.* 1.

Genus [10]. LAMPRIAS. *Bonelli.* Four species.

L. chlorocephalus, *Ent. Heft.* Rich bluish green, bright and shining, antennæ pitchy, except their base, which with the thorax and legs (except the tarsi which are black) testaceous-red ; two to three lines ; broom. *Pl.* 2, *fig.* 2.

Family [3]. SCARITIDÆ. Macleay.

Genus [11]. SCARITES. *Fab.* One species.

S. Beckwithii, *Steph.* Entirely black ; elytra striated ; eight and a half lines ; very rare ; coasts. *Pl.* 2, *fig.* 3.

Genus [12]. CLIVINA. *Lat.* Two species.

C. fossor, *Lin.* Pitchy brown ; antennæ and legs red ; elytra punctato-striate ; three and a half lines ; humid situations. *Pl.* 2, *fig.* 4.

Genus [13]. DYSCHIRIUS. *Bonelli.* Sixteen species.

C. nitidus, *Dej.* Bronzy, shining, very smooth and convex ; tibiæ ferruginous ; two lines ; marshy pastures. *Pl.* 2, *fig.* 5.

Family [4]. HARPALIDÆ. Macleay.

GROUP I. HARPALINI. ERICHSON.

Genus [14]. SELENOPHORUS. *Dej.* One species.

C. Scaritides, *Zieg.* Black and shining ; antennæ and tarsi pitchy red ; elytra striated ; four lines ; banks of Thames, Gravesend. *Pl.* 2, *fig.* 6.

Genus [15]. ANISODACTYLUS. *Dej.* Two species.

A. binotatus, *Fab.* Black ; two basal joints of antennæ ferruginous, tarsi pitchy red ; five lines ; under stones. *Pl.* 2, *fig.* 7.

Genus [16]. DIACHROMUS. *Erichs.* One species.

D. Germanus, *Lin.* Black, head, legs and elytra testaceous, the latter with a large blue spot at the apex ; thorax with a bluish tinge ; four lines ; Devonshire, rare. *Pl.* 2, *fig.* 8.

Genus [17]. HARPALUS. *Latr.* Fifty-one species.

H. ruficornis, *Fab.* Pitchy black ; legs and antennæ red ; elytra pubescent ; five to seven lines ; common. *Pl.* 3, *fig.* 1.

Genus [18]. ACTEPHILUS. *Steph.* Two species.

A. pumilus, *Stur.* Pitchy black ; antennæ at the base and the palpi reddish ; three and a half lines ; banks of Thames, rare. *Pl.* 3, *fig.* 2.

Genus [19]. OPHONUS. *Zieg.* Differs chiefly from Harpalus, by its densely punctured surface. Fifteen species.

O. obscurus, *Fab.* Black ; elytra with a bluish tinge ; antennæ and legs ferruginous ; 5—6 lines ; under stones, humid situations. *Pl.* 2, *fig.* 3.

Genus [20]. STENOLOPHUS. *Dej.* Four species.

S. vaporariorum, *Fab.* Pitchy black shining ; two basal joints of antennæ ; thorax ; humeral angles of elytra and legs testaceous ; two and a half to three and a half lines ; humid situations. *Pl.* 3, *fig.* 4.

Genus [21]. MASOREUS. *Dej.* One species.

M. Wetterhallii, *Gyll.* Pitchy black, shining; antennæ and legs red; base of elytra, reddish brown; two and a half lines; under stones in estuaries. *Pl. 3, fig. 5.*

GROUP II. PÆCILINI. SHUCKARD.

Genus [22]. POGONUS. *Zieg.* Four species.
  P. Burrellii, *Curt.* Greenish bronzy; antennæ fuscescent; elytra ochraceous; legs testaceous; four lines; coasts of Norfolk. *Pl. 3, fig. 6.*
Genus [23]. PÆCILUS. *Bonelli.* Four species.
  P. dimidiatus, *Fab.* Black; head and thorax with a rich coppery-red tinge; elytra silky green; six lines; sandy heaths. *Pl. 3, fig. 7.*
Genus [24]. Sogines punctulatus is not British and therefore not figured here.
Genus [25]. OMASEUS. *Zieg.* Eleven species.
  O. melanarius, *Illig.* Atrous black and shining; 7—9½ lines; common beneath stones in gardens. *Pl. 3, fig. 8.*
Genus [26]. ARGUTOR. *Megerle.* Nine species.
  A. inequalis, *Marsh.* Atrous black and shining; three lines; under stones on the banks of the Thames. *Pl. 4, fig. 1.*
Genus [27]. PLATYDERUS. *Steph.* One species.
  P. ruficollis, *Marsh.* Pitchy red, shining; thorax and legs rather paler; 3—4 lines; under moss, roots of trees. *Pl. 4, fig. 2.*
Genus [28]. STEROPUS. *Megerle.* Four species.
  S. madidus, *Fab.* Pitchy black, shining; femora either red or black; 6½—8 lines; common. *Pl. 4, fig. 3.*
Genus [29]. BROSCUS. *Panz.* One species.
  B. cephalotes, *Lin.* Atrous; head and thorax shining; 8½—10¼ lines; sandy coasts. *Pl. 4, fig. 4.*
Genus [30]. MISCODERA. *Esch.* One species.
  M. arctica, *Payk.* Æneous, very shining and very convex; legs red; three and a half lines; north of England. *Pl. 4, fig. 5.*
Genus [31]. STOMIS. *Clairville.* One species.
  S. pumicatus, *Panz.* Pitchy black; antennæ and legs red; three and a half lines; under leaves and felled timber. *Pl. 4, fig. 6.*
Genus [32]. PATROBUS. *Megerle.* Two species.
  P. excavatus, *Payk.* Pitchy black; legs reddish; four lines; coasts. *Pl. 4, fig. 7.*
Genus [33]. PTEROSTICHUS. *Bonelli.* One species.
  P. parumpunctatus, *Germ.* Atrous, shining; apex of anten-

næ pitchy; seven and a half lines; Newcastle, under stones. *Pl.* 4, *fig.* 8.

Genus [34]. ADELOSIA. *Steph.* Two species.

    A. oblongopunctata, *Fab.* Pitchy black, very depressed; legs pitchy red; six lines; humid situations. *Pl.* 5, *fig.* 1.

Genus [35]. PLATYSMA. *Bonelli.* One species.

    P. nigrum, *Fab.* Deep black; 9—11 lines, gardens and woods. *Pl.* 5, *fig.* 2.

Genus [36]. ABAX. *Bon.* One species.

    A. striola, *Fab.* Deep black, shining in the male, elytra obscure in the female; $8\frac{1}{2}$—10 lines; under stones. *Pl.* 5, *fig.* 3.

GROUP III. AMARINI. SHUCKARD.

Genus [37]. AMARA. *Bon.* Eighteen species.

    A. acuminata, *Payk.* Æneous, coppery, or black; base of antennæ fulvous; 4—6 lines; pastures and fields. *Pl.* 5, *fig.* 4.

Genus [38]. CELIA. *Erichs.* Differs from Amara by the mentum tooth being emarginate, and the posterior tibiæ of the male glabrous. Three species.

    C. ingenua, *Duft.* Fusco-æneous; antennæ and legs red; four lines; rare. *Pl.* 5, *fig.* 5.

Genus [39]. ACRODON. *Zimmerm.* Differs from Amara chiefly by the mentum tooth being very acute, and the posterior tibiæ of the male only slightly pilose. One species.

    A. brunnea, *Gyll.* Fusco-æneous; antennæ and legs red; two and a half lines; in woods. *Pl.* 5, *fig.* 6.

Genus [40]. BRADYTUS. *Steph.* Seven species.

    B. fulvus, *De Geer.* Entirely fulvous; elytra with an opaline tinge; four lines; sandy situations. *Pl.* 5, *fig.* 7.

Genus [41]. CURTONOTUS. *Steph.* Three species.

    C. convexiusculus, *Marsh.* Pale pitchy; antennæ and legs fulvous; elytra slightly æneous; five and a half lines; shores on the coasts. *Pl.* 5, *fig.* 8.

Genus [42]. ZABRUS. *Clairville.* One species.

    L. gibbus, *Fab.* Black, very convex; antennæ and tibiæ pitchy; 6—$7\frac{1}{2}$ lines; corn fields. *Pl.* 6, *fig.* 1.

GROUP IV. ANCHOMENINI. ERICHSON.

Genus [43]. CALATHUS. *Bonelli.* Eleven species.

    C. Cisteloides, *Lin.* Entirely black; 5—$6\frac{1}{2}$ lines; common. *Pl.* 6, *fig.* 2.

Genus [44]. ODONTONYX. *Steph.* One species.

O. rotundicollis, *Marsh.* Pitchy æneous ; base of antennæ
and legs testaceous ; three and a half lines; fields. *Pl.* 6,
fig. 3.

Genus [45]. TAPHRIA. *Bonelli.* One species.

T. vivalis, *Panz.* Black ; antennæ and legs red ; four lines;
humid woods. *Pl.* 6, *fig.* 4.

Genus [46]. PRISTONYCHUS, *Dej.* Differs from Sphodrus by its
claws being serrated at the base within, and posterior trochanters of the male not produced. One species.

P. subcyaneus, *Illig.* Pitchy black; elytra with a violet tinge ;
antennæ and legs sometimes paler ; 7—8 lines ; damp cellars. *Pl.* 6, *fig.* 5.

Genus [47]. SPHODRUS. *Claire.* Differs from Pristonychus by
the claws being simple and not serrated, and the posterior
trochanters of the male produced into a long spine. One
species.

S. leucophthalmus, *Lin.* Opaque black ; 10—11½ lines ;
damp cellars. *Pl.* 6, *fig.* 6.

Genus [48]. PLATYNUS. *Bonelli.* One species.

P. angusticollis, *Fab.* Entirely of a bright shining black ;
six lines ; woods, roots of trees. *Pl.* 6, *fig.* 7.

Genus [49]. AGONUM. *Bonelli.* Thirty species.

A. marginatum, *Lin.* Coppery green or reddish above ;
margin of elytra and tibiæ testaceous ; 4—5 lines ; banks
of ponds and marshy places, common. *Pl.* 6, *fig.* 8.

Genus [50]. OLISTHOPUS. *Dej.* One species.

O. rotundatus, *Payk.* Fusco-æneus ; antennæ and legs testaceous ; three and a half lines ; corn fields. *Pl.* 7,
fig. 1.

Genus [51]. ANCHOMENUS. *Bonelli.* Three species.

A. prasinus, *Fab.* Red ; head, thorax, and a large common
patch, at the apex of the elytra, green ; antennæ fuscous
from the base ; three and a half lines ; fields. *Pl.* 7,
fig. 2.

GROUP V. TRECHINI. SHUCK.

Genus [52]. AËPUS. *Leach.* One species.

A. fulvescens, *Leach.* Entirely pale testaceous ; flat and
shining ; one line ; under stones below high water mark.
Pl. 7, *fig.* 3.

Genus [53]. EPAPHIUS. *Leach.* One species.

E. secalis, *Payk.* Ferrugineous ; shining ; antennæ fuscous,
legs testaceous ; eastern coasts. *Pl.* 7, *fig.* 4.

Genus [54]. BLEMUS. *Ziegl.* Six species.

B. discus, *Fab.* Red testaceous; legs paler; elytra with a dark cloud across their centre ; two and a half lines ; roads and gardens. *Pl. 7, fig. 5.*

Genus [55]. BRADYCELLUS. *Erichson.* Six species.
B. dorsalis, *Lin.* Testaceous; head black ; and a large common patch on the elytra pitchy ; antennæ, except the base, fuscous ; two lines ; humid situations. *Pl. 7, fig. 6.*

Genus [56]. TRECHUS. *Clairville.* Eleven species.
T. meridianus, *Lin.* Pitchy black ; the base and suture of the elytra as well as the legs testaceous ; one and a half lines ; marshy places. *Pl. 7, fig. 7.*

GROUP VI. LICININI. ERICH.

Genus [57]. CALLISTUS. *Bonelli.* One species.
C. lunatus, *Fab.* Black head with a greenish tinge ; thorax red ; elytra with a double testaceous cross ; legs also testaceous, except the knees of the femora which are black ; three and a half lines; chalky downs. *Pl. 8, fig. 1.*

Genus [58]. OODES. *Bonelli.* One species.
O. helopoides, *Fab.* Entirely black ; four and a half lines ; marshy situations. *Pl. 8, fig. 2.*

Genus [59]. CHLÆNIUS. *Bonelli.* Six species.
C. vestitus, *Fab.* Greenish coppery above ; antennæ, margin and apex of elytra and legs testaceous ; 5—6 lines ; humid situations. *Pl. 8, fig. 3.*

Genus [60]. EPOMIO. *Bonelli.* One species.
E. circumscriptus, *Duft.* Head and thorax with a green reflection ; elytra black, their margin and the legs testaceous; eight lines . . . . ? a very doubtful native. *Pl. 8, fig. 4.*

Genus [61]. LICINUS. *Latr.* Two species.
L. silphoides, *Fab.* Entirely obscure black ; 6—7 lines ; chalky downs. *Pl. 8, fig. 5.*

Genus [62]. BADISTER. *Clairville.* Three species.
B. bipustulatus, *Fab.* Red-testaceous ; very shiny, head, middle of antennæ and a waved band constricted at the suture crossing the elytra just beyond the middle : all black ; elytra with an opaline reflection ; three and a half lines ; marshy places. *Pl. 9, fig. 6.*

Genus [63]. TRIMORPHUS. *Steph.* Three species.
T. scapularis, *Steph.* Pitchy black ; narrow edge of thorax, humeral angles of elytra and legs testaceous ; 3—4 lines ; woody places, roots of trees. *Pl. 8, fig. 7.*

Genus [64]. PANAGŒUS. *Latr.* Two species.
P. crux-major, *Lin.* Deep black ; elytra red with a cross and the apex black ; four lines ; woods and fens. *Pl. 9, fig. 1.*

8

Genus [65]. LORICERA. *Latr.* One species.
    L. pilicornis, *Fab.* Bronzy above; tibiæ and tarsi piceous; three and a half lines; about roots of trees. *Pl.* 9, *fig.* 2.

Family [5]. CARABIDÆ. Macleay.

Genus [66]. LEISTUS. *Fröl.* Eight species.
    L. spinibarbis, *Fab.* Steely blue above; mouth, antennæ, tibiæ and tarsi pitchy; 4—5 lines; under stones and sticks in hedges. *Pl.* 9, *fig.* 3.
Genus [67]. HELOBIA. *Leach.* Seven species.
    H. brevicollis, *Fab.* Black; legs and antennæ pitchy; 5—6 lines; common. *Pl.* 9, *fig.* 4.
Genus [68]. NEBRIA. *Latr.* Three species.
    N. complanata, *Lin.* Testaceous; elytra maculated with black, usually in two broad irregular, laterally abbreviated bands; the colour frequently suffused; 8—9 lines; shores, especially those of the British Channel. *Pl.* 9, *fig.* 5.
Genus [69]. PELOPHILA. *Dej.* One species.
    P. Borealis, *Fab.* Æneous; legs red; seven lines; Ireland. *Pl.* 9, *fig.* 6.
Genus [70]. CALOSOMA. *Fab.* Two species.
    C. inquisitor, *Lin.* Coppery or æneous above; margin of elytra green; legs and antennæ black; 8—10 lines; trees, Hainault forest. *Pl.* 10, *fig.* 1.
Genus [71]. CARABUS. *Lin.* Sixteen species.
    C. catenulatus, *Fab.* Black; the thorax and margins of elytra violaceous; 10—12 lines; sandy places, common. *Pl.* 10, *fig.* 2.
Genus [72]. CYCHRUS. *Fab.* One species.
    C. rostratus, *Lin.* Entirely black; nine lines; hollow trees and under leaves in woods. *Pl.* 10, *fig.* 3.

Family [6]. ELAPHRIDÆ. Stephens.

Genus [73]. ELAPHRUS. *Fab.* Four species.
    E. cupreus, *Meg.* Entirely bronzy; four lines; humid places. *Pl.* 10, *fig.* 4.
Genus [74]. BLETHISA. *Bonelli.* One species.
    B. multi-punctata, *Lin.* Bronzy; very shining; legs black; six lines; marshes. *Pl.* 10, *fig.* 5.
Genus [75]. NOTIOPHILUS. *Dumeril.* Eighteen species?
    N. aquaticus, *Lin.* Æneous, brilliant shining; two and a half lines; marshes. *Pl.* 10, *fig.* 6.

Family [7]. BEMBIDIIDÆ. Stephens.

Genus [76]. BEMBIDIUM. *Latr.* Four species.
    B. paludosum, *Panz.* Entirely obscure bronzy; $2\frac{1}{2}$ lines; fenny situations. *Pl.* 11, *fig.* 1.
Genus [77]. TACHYPUS. *Megerle.* Eight species.
    T. pallidipennis, *Illig.* Head and thorax cupreus; legs and elytra testaceous, the latter with a fuscous angulated band, just beyond the middle; two lines; coasts and estuaries. *Pl.* 11, *fig.* 2.
Genus [78]. NOTAPHUS. *Megerle.* Ten species.
    N. undulatus, *Sturm.* Head and thorax blackish-æneous; antennæ, legs and elytra brown; the latter with a waved pale fascia towards the apex; $2\frac{1}{2}$—3 lines; humid situations. *Pl.* 11, *fig.* 3.
Genus [79]. LOPHA. *Megerle.* Fourteen species.
    L. quadriguttata, *Fab.* Blackish-æneous, shining; the elytra with four pale spots, two humeral and two just beyond the middle, placed externally; tibiæ and tarsi testaceous, terminal joints of the latter black; $2\frac{1}{4}$ lines; humid places. *Pl.* 11, *fig.* 4.
Genus [80]. PERYPHUS. *Megerle.* Twenty-two species.
    C. concinnus, *Kirb.* Head and thorax æneous; base of antennæ, legs and elytra testaceous, the latter with a central diamond shaped black mark; $2\frac{1}{2}$ lines; humid situations. *Pl.* 11, *fig.* 5.
Genus [81]. OCYS. *Kirby.* Three species.
    O. tempestivus. *Panz.* Ferruginous, head and thorax rather paler; antennæ and legs testaceous; $2\frac{1}{4}$ lines; beneath the bark of trees. *Pl.* 11, *fig.* 6.
Genus [82]. PHILOCHTHUS. *Stephens.* Six species.
    P. Doris, *Marsh.* fuscipes, *Dej.* Blackish-æneous; elytra with the apex pale; legs ferruginous; two lines; humid places. *Pl.* 11, *fig.* 7.
Genus [83]. TACHYS. *Meg.* Nine species.
    T. obtusus, *Stur.* Pitchy black; antennæ fuscous; their base and the legs pitchy red; $1\frac{1}{2}$ line; banks of ponds, common. *Pl.* 12, *fig.* 1.
Genus [84]. CILLENUM. *Leach.* One species.
    C. laterale, *Leach.* Coppery, shining; antennæ fuscous; their base and the legs testaceous; elytra ochraceous, with an æneous reflection; two lines; coasts under stones. *Pl.* 12, *fig.* 2.
Genus [85]. LYMNÆUM. *Steph.* Three species.
    L. depressum, *Curt.* Pitchy black; antennæ fuscous; with

C

their base and the legs fulvous; 1¼ line; coasts, Lanca-
shire. *Pl. 12, fig. 3.*

TRIBE II. HYDRADEPHAGA. Macleay.

Race I. EUNECHINA. Kirb. and Sp.

Family [8]. Dytiscidæ. Leach.

Subfamily I. HALIPLITES. Shuckard.

Group I. HALIPLINI. Erichson.

Genus [86]. Haliplus. *Latr.* Differs from Cnemidotus, by
the terminal joint of palpi being subulated, and the plate of
the coxæ leaving three segments of the venter exposed. Fifteen species.

    H. ferruginous, *Lin.* Entirely ferruginous-red; the legs
    slightly paler; 2—2½ lines; ditches and ponds, common.
    Pl. 12, fig. 4.

Genus [87]. Cnemidotus. *Illig.* Differs from Haliplus, by
the terminal joint of palpi being conic acute, and the plate of
the coxæ leaving only one segment of the venter exposed.
One species.

    C. cæsus, *Duft.* Yellow-testaceous; the elytra with suc-
    cessive rows of black punctures, which decrease in size to
    the apex; legs fulvous; about two lines; ponds and
    ditches. *Pl. 12, fig. 5.*

Group II. PELOBINI. Erichson.

Genus [88]. Pelobius. *Schön.* One species.

    P. Hermanni, *Fab.* Fulvous; thorax with the base and
    apex, and the elytra with a large black irregular common
    patch on the disk, red; legs testaceous; five lines; ponds
    and ditches. *Pl. 12, fig. 6.*

Subfamily II. DYTISCITES. Erichson.

Group I. HYDROPORINI. Erichson.

Genus [89]. Hyphydrus. *Illig.* One species.

    H. ovatus, *Lin.* Pale ferruginous; the elytra more obscure;
    two lines; ponds and ditches, common. *Pl. 12, fig. 7.*

Genus [90]. HYGROTUS. *Steph.* Teu species.
 H. confluens. *Fab.* Testaceous; head slightly darker, with
the occiput black or dusky; elytra with five black, longitudinal, parallel lines, the first and third from the suture
abbreviated in front, and all confluent behind; $1\frac{3}{4}$ line;
common. *Pl.* 13, *fig.* 1.
Genus [91]. HYDROPORUS. *Clairv.* Forty-four species.
 H. 12. pustulatus, *Fab.* Ochraceous, the thorax with a
basal bilobate, black mark; elytra also black and each
with six pale spots, three sutural and three marginal;
three lines; ponds and ditches. *Pl.* 13, *fig.* 2.

GROUP II. COLYMBETINI. ERICHSON.

Genus [92]. NOTERUS. *Clairv.* Two species.
 N. crassicornis, *Fab.* Ferruginous; elytra darker, with
their margin pale; $2\frac{1}{2}$ lines; ponds and ditches. *Pl.* 13,
fig. 3.
Genus [93]. LACCOPHILUS. *Leach.* Two species.
 L. minutus, *Lin.* Testaceous-yellow; the elytra fuscous
with the margin and some irregular discoidal spots, paler;
two lines; ponds and ditches. *Pl.* 13, *fig.* 4.
Genus [94]. AGABUS. *Leach.* Twenty-nine species. In the
type only are the antennæ in the male serrated; in the female
the antennæ and feet are always simple. The genus is also di-
vided into sections and subsections from the acetabula of the
underside of the male anterior tarsi and the ciliation of their
posterior tarsi.
 A. serricornis, *Payk.* Pitchy; the lateral margins of the
thorax and elytra rusty red; six lines; very rare, Shrop-
shire. *Pl.* 13, *fig.* 5.
Genus [95]. ILYBIUS. *Erichs.* Six species.
 I. ater, *Fab.* Black, subæneous, the external margins reddish brown; the elytra with two pale pellucid spots; six
lines; common. *Pl.* 13, *fig.* 6.
Genus [96]. COLYMBETES. *Clairv.* Six species. Anterior
tarsi of male with three dilated joints.
 C. striatus, *Lin.* Fuscous; thorax ferruginous, black in the
centre; margins of elytra testaceous and legs black; eight
lines; ponds and ditches, common. *Pl.* 14, *fig.* 1.

GROUP III. DYTISCINI. ERICHSON.

Genus [97]. DYTISCUS. *Lin.* Five species. In the female
anterior tarsi simple, and elytra usually furrowed longitudinally.

12

D. circumflexus, *Fab.* Olivaceous; margins of thorax, lateral margins of elytra, and legs externally, testaceous; posterior tarsi black; one inch and four lines; ponds, London, common. *Pl.* 14, *fig.* 2.

Genus [98]. HYDATICUS. *Leach.* Four species. Tarsi in the female simple.

H. transversalis, *Fab.* Pitchy black; with the face in front, thorax, excepting a patch at the base, lateral margins, and an abbreviated, transverse, sinuated line at the base of the suture all ferruginous; legs pitchy red; 2½ lines; ponds. *Pl.* 14, *fig.* 3.

Genus [99]. ACILIUS. *Leach.* Two species. In female anterior tarsi simple, and elytra longitudinally furrowed.

A. sulcatus, *Lin.* Fuscous; the mouth and the margins, and central transverse fascia of the thorax yellowish; legs testaceous; their tarsi ferruginous; 6—9 lines; very common; ponds. *Pl.* 14, *fig.* 4.

Genus [100]. CYBISTER. *Curt.* One species. Tarsi in female simple.

C. Roeselii, *Fab.* Olive black, face in front, and lateral margins of thorax and elytra; dirty testaceous; legs testaceous; posterior tibiæ and tarsi pitchy; one inch and one line; ponds, very rare. *Pl.* 14, *fig.* 5.

RACE II. GYRONECHINA. KIRB. AND SP.

Family [9]. GYRINIDÆ. Leach.

Genus [101]. GYRINUS. *Lin.* Eight species.

G. natator, *Lin.* Glossy blue black; mouth rufescent; legs pale ferruginous; three lines; ponds and ditches, common. *Pl.* 13, *fig.* 7.

Genus [102]. ORECTOCHILUS. *Esch.* One species. In female the tarsi simple.

O. villosus, *Fab.* Olivaceous, villose; antennæ ferruginous; legs testaceous; three lines, running streams, uncommon. *Pl.* 13, *fig.* 8.

SUBDIVISION II. BRACHELYTRA. LATR.

Family [10]. OMALIDÆ. Macleay.

Genus [103]. LESTEVA. *Latr.* Fourteen species.

L. caraboides, *Lin.* Reddish testaceous ; elytra and legs pale testaceous ; two lines ; northern counties. *Pl.* 15, *fig.* 1.

Genus [104]. CORYPHIUM. *Kirb.* One species.

C. angusticolle, *Kirb.* Black, slightly shining ; the base of the antennæ and legs testaceous ; 1¼ line ; meadows by sweeping. *Pl.* 15, *fig.* 2.

Genus [105]. ACIDOTA. *Kirb.* Two species.

A. crenata, *Fab.* Black ; sides of thorax, elytra and legs brown ; three lines ; Scotland. *Pl.* 15, *fig.* 3.

Genus [106]. OMALIUM. *Grav.* Thirty species.

O. planum, *Payk.* Black ; base of antennæ, elytra and legs testaceous-brown ; 1¼ line ; meadows by sweeping. *Pl.* 15, *fig.* 4.

Genus [107]. MICRALYMMA. *Westw.* One species.

M. Johnstonis, *Westw.* Entirely black ; 1¼ line ; Scotland. *Pl.* 15, *fig.* 5.

Genus [108]. ANTHOBIUM. *Leach.* Twenty-one species.

A. melanocephalum, *Marsh.* Yellow testaceous ; head dusky ; 1¼ line ; meadows by sweeping. *Pl.* 15, *fig.* 6.

Genus [109]. SYNTOMIUM. *Curt.* One species.

S. nigro-æneum, *Curt.* Blackish-æneous ; legs and antennæ pitchy ; not quite a line long ; grassy places by sweeping. *Pl.* 15, *fig.* 7.

Genus [110]. PROTEINUS. *Latr.* Three species.

P. brachypterus, *Payk.* Atrous ; basal joint of antennæ, palpi and legs testaceous ; not quite a line long; putrid fungi. *Pl.* 15, *fig.* 8.

Genus [111]. MEGARTHRUS. *Kirb.* Seven species.

M. rufescens, *Kirb.* Reddish pitchy ; antennæ, lateral margins of thorax and legs red ; abdomen black ; about a line long ; putrid fungi. *Pl.* 16, *fig.* 1.

Genus [112]. PSEUDOPSIS. *Newm.* One species.

P. sulcatus, *Newm.* Black ; mouth, antennæ and legs fuscous ; 1½ line ; Isle of Wight. *Pl.* 16, *fig.* 2.

Genus [113]. COPROPHILUS. *Latr.* One species.

C. striatulus, *Fab.* Black ; legs and antennæ pitchy ; 2½ lines ; gardens, roads, and meadows. *Pl.* 16, *fig.* 3.

Genus [114]. TÆNOSOMA. *Mannerheim.* One species.

T. pusillum, *Knoch.* Black ; antennæ, elytra and legs testaceous ; two thirds of a line ; rare. *Pl.* 16, *fig.* 4.

Genus [115]. PHLŒOCHARIS. *Mann.* One species.

P. subtilissima, *Mann.* Pitchy black, pubescent ; antennæ and legs fuscous ; one line long ; beneath bark ; rare. *Pl.* 16, *fig.* 5.

Family [11]. OXYTELIDÆ. Shuck.

Genus [116]. TROGOPHLŒUS. *Mann.* Nine species.
> T. corticinus, *Grav.* Pitchy black ; base of antennæ, tibiæ and tarsi red; 1¼ line ; beneath bark ; rare. *Pl.* 16, *fig.* 6.

Genus [117]. APLODERUS. *Steph.* One species.
> A. brachypterus, *Mann.* Black; antennæ, elytra and apex of abdomen reddish ; legs testaceous ; 2¼ lines ; meadows by sweeping. *Pl.* 16, *fig.* 7.

Genus [118]. OXYTELUS. *Grav.* Sixteen species.
> O. carinatus, *Grav.* Black ; tibiæ and tarsi testaceous ; nearly two lines ; dung of horses ; very common. *Pl.* 16, *fig.* 8.

Genus [119]. PLATYSTETHUS. *Mann.* Nine species. Head and thorax smaller and less robust in the female.
> P. morsitans, *Payk.* Black and shining; elytra and legs pitchy; 1½ line ; spring, dung of horses ; common. *Pl.* 17, *fig.* 1.

Genus [120]. PHYTOSUS. *Rudd.* One species. Elytra longer and wider than the thorax in the female.
> P. spinifer, *Rudd.* Male. Head and abdomen, except its apex, pitchy black; antennæ, thorax, elytra, legs and apex of abdomen reddish. Female, black, with the mouth, antennæ and legs pitchy; ¾—1¼ line ; beneath marine rejectamenta ; Ryde, Isle of Wight. *Pl.* 17, *fig.* 2.

Genus [121]. HESPEROPHILUS. *Steph.* Five species.
> H. fracticornis, *Payk.* Black ; base of antennæ and legs reddish ; 2½ lines ; sandy coasts. *Pl.* 17, *fig.* 3.

Genus [122]. BLEDIUS. *Leach.* Three species. Thorax unarmed in the female.
> B. tricornis, *Payk.* Black; elytra and apex of abdomen castaneous and legs pitchy ; three lines ; banks of ponds and brooks. *Pl.* 17, *fig.* 4.

Family [12]. STENIDÆ. Macleay.

Genus [123]. STENUS. *Latr.* Sixty-four species.
> S. biguttatus, *Lin.* Black, shining, with an æneous reflection, the elytra having each a central fulvous spot ; 2½ lines ; banks of ponds. *Pl.* 17, *fig.* 5.

Genus [124]. DIANOUS. *Leach.* One species.
> D. cærulescens, *Gyll.* Black, shining, with a bluish tinge, each of the elytra with a central fulvous spot ; 2½ lines ; humid situations in damp moss. *Pl.* 17, *fig.* 6.

Genus [125]. PÆDERUS. *Fab.*

P. riparius, *Lin.* Black; elytra bluish; thorax, four first segments of abdomen, mesosternum and legs red; knees of the femora and the antennæ black, the base of the latter testaceous; 3½ lines; humid situations. *Pl.* 17, *fig.* 7.

Genus [126]. RUGILUS. *Leach.* (STILICUS, *Latr.*) Four species.

R. orbiculatus, *Payk.* Black, thorax with a central longitudinal carina; elytra with the apex pitchy and the legs testaceous; two lines; by sweeping; not uncommon. *Pl.* 17, *fig.* 8.

Genus [127]. ASTENUS. *Dej.* Five species.

A. angustatus, *Fab.* Black; with the mouth, antennæ, legs and apex of the elytra testaceous; 1¾ line; by sweeping humid meadows; not uncommon. *Pl.* 18, *fig.* 1.

Genus [128]. SUNIUS. *Leach.* Four species.

S. melanocephalus, *Fab.* Black, much punctured; antennæ, thorax and legs fulvous; 1¾ line; beneath stones on heaths. *Pl.* 18, *fig.* 2.

Genus [129]. EVÆSTHETUS. *Grav.* One species.

E. scaber, *Grav.* Pitchy black opaque, head and legs red; one line; beneath bark. *Pl.* 18, *fig.* 3.

Genus [130]. MEDON. *Steph.* (LITHOCARIS, *Erichs.*) Two species.

M. Ruddii, *Steph.* Black and shining; antennæ, mouth, elytra, legs and apex of the abdomen red; 3¼ lines; New Forest. *Pl.* 18, *fig.* 4.

The second species hitherto unrecorded is M. ochraceus, *Erichs.*

Family [13]. STAPHYLINIDÆ. Leach.

Genus [131]. SIAGONIUM. *Kirby and Spence.* One species.
In the female the head is narrower and not cornuted.

S. quadricorne, *K. and S.* Pitchy black shining, antennæ, legs, an oblique mark on the elytra, and the apex of the abdomen reddish: or sometimes, entirely reddish; 2—2½ lines; under the bark of elms. *Pl.* 18, *fig.* 5.

Genus [132]. ACHENIUM. *Leach.* One species.

A. depressum, *Grav.* Very depressed, black; antennæ pitchy; legs and terminal half of the elytra red; 3½ lines; banks of ponds, Battersea. *Pl.* 18, *fig.* 6.

Genus [133]. CRYPTOBIUM. *Mann.* One species.

C. fracticorne, *Payk.* Black and shining; legs testaceous; 2½ lines; Battersea fields, rare. *Pl.* 18, *fig.* 7.

16

Genus [134]. LATHROBIUM. *Grav.* Fourteen species.
  L. elongatum, *Lin.* Black; antennæ, legs and terminal half of elytra, pitchy red; four lines; common. *Pl.* 18, *fig.* 8.
Genus [135]. GYROHYPNUS. *Kirby.* Twenty species.
  G. cruentatus, *Marsh.* Atrous shining; elytra red; six lines; common. *Pl.* 19, *fig.* 1.
Genus [136]. OTHIUS. *Leach.* Eleven species.
  O. fulgidus, *Payk.* Black; antennæ, elytra, apex of abdomen and legs red; five lines; gardens and roads. *Pl.* 19, *fig.* 2.
Genus [137]. GABRIUS. *Leach.* Thirteen species.
  G. suaveolens, *Kirby.* Black and shining; elytra opaque; antennæ pitchy; their base as well as the palpi and legs testaceous; 2½ lines; beneath rejectamenta, Norfolk. *Pl.* 19, *fig.* 3.
Genus [138]. CAFIUS. *Leach.* Five species.
  C. fucicola, *Leach.* Opaque black; legs slightly pitchy; two to four lines; beneath marine rejectamenta. *Pl.* 19, *fig.* 4.
Genus [139]. BISNIUS. *Leach.* Five species.
  B. cephalotes, *Grav.* Black; elytra æneous; thorax and legs pitchy; 2¾ lines; Norfolk. *Pl.* 19, *fig.* 5.
Genus [140]. HETEROTHOPS. *Kirby.* Three species.
  H. binotatus, *Kirby.* Black and shining; base of antennæ, and legs pitchy; 2¼ lines; coasts, beneath marine rejectamenta. *Pl.* 19, *fig.* 6.
Genus [141]. RAPHIRUS. *Leach.* Thirteen species.
  R. semiobscurus, *Marsh.* Black, head and thorax shining; antennæ and legs fulvous; 3¼ lines; beneath rejectamenta. *Pl.* 19, *fig.* 7.
Genus [142]. PHILONTHUS. *Leach.* Forty-eight species.
  P. politus, *Lin.* Blackish-æneous; abdomen entirely black; head and thorax shining; 3½—5 lines; very common. *Pl.* 19, *fig.* 8.
Genus [143]. QUEDIUS. *Leach.* Thirty-eight species. Differs from Euryporus, *Erichs.* by the palpi being filiform.
  Q. tristis, *Grav.* Pitchy black; antennæ and legs pitchy red; 5½ lines; under stones, common. *Pl.* 20, *fig.* 1.
Genus [144]. ASTRAPÆUS. *Grav.* One species. Has the terminal joint of all the palpi securiform.
  A. Ulmi, *Rossi.* Black, with the base of the antennæ, labrum, margin of the clypeus, elytra and penultimate segment of the abdomen red; tibiæ and tarsi pitchy; five lines; beneath the bark of elms, rare. *Pl.* 20, *fig.* 2.

Genus [145]. OXYPORUS. *Fab.* Two species. Has the terminal joint of maxillary palpi ovate, and of the labial lunate.

    O. rufus, *Lin.* Red; with the head, elytra, except their humeral angles, base of the femora and apex of the abdomen black; four lines; in boleti in the autumn; not uncommon. *Pl. 20, fig. 3.*

Genus [146]. TASGIUS. *Leach.* Two species. Has the terminal joint of labial palpi only, securiform.

    T. rufipes, *Latr.* Pitchy opaque black; with the base and apex of the antennæ and legs red; seven lines; in sand pits; Hampstead, rare. *Pl. 20, fig. 4.*

Genus [147]. OCYPUS. *Kirb.* Five species. Has the mandibles edentate and terminal joint of all the palpi truncated.

    O. similis, *Oliv.* Pitchy black opaque; head and thorax with an æneous reflection, and anterior tarsi, reddish, pitchy; seven lines; sand pits on heaths; common. *Pl. 20, fig. 5.*

Genus [148]. EURYPORUS. *Erichs.* (Pelecyphorus, *Nordm.*) One species. Differs from *Quedius* by the terminal joint of the labial palpi being securiform, and from *Tasgius* by the terminal joint of the maxillary palpi being acuminated.

    E. picipes, *Gyll.* Black and shining; the mouth and base of the antennæ red; legs pitchy red; beneath the bark of trees; rare. *Pl. 20, fig. 6.*

Genus [149]. GOERIUS. *Leach.* Eight species.

    G. olens, *Müll.* Entirely black, opaque and pubescent; 8—15 lines; common; gardens and pathways. *Pl. 20, fig. 7.*

Genus [150]. STAPHYLINUS. *Linn.* Twelve species.

    S. erythropterus, *Linn.* Black; with the base of the antennæ, elytra and legs reddish-testaceous; a patch on each side of the head above the insertion of the antennæ, posterior margin of the thorax, and of the first segment of the abdomen, and a triangular spot on each side, at the base of the three following, of a rich golden pubescence; 6—9 lines; common; sandy situations. *Pl. 20, fig. 8.*

Genus [151]. EMUS. *Leach.* One species.

    E. hirtus, *Lin.* Black pubescent; the head, thorax and apex of the abdomen covered with long, bright, yellow hair; the elytra with a silvery ashy band behind; 8—12 lines; heaths in Hampshire; not common. *Pl. 21, fig. 1.*

Genus [152]. CREOPHILUS. *Kirb.* Two species?

    C. maxillosus, *Lin.* Deep black; head and thorax glabrous and shining; elytra and abdomen pubescent, the former with an ashy band, and the abdomen with the third and

D

fourth segments laterally ashy, and all upon the disk with a waved band of the same colour; 6—12 lines; common. *Pl.* 21, *fig.* 2.

Genus [153]. VELLEIUS. *Leach.* One species.

    V. dilatatus, *Fab.* Black; head and thorax shining, the latter having an æneous reflection; elytra and abdomen opaque and slightly pubescent; 9 lines; Hornet's nests? rare. *Pl.* 21, *fig.* 3.

Family [14]. TACHINIDÆ. Shuck.

Genus [154]. TACHINUS. *Grav.* Twenty-three species.

    T. latus, *Marsh.* Black and shining; base of antennæ and legs pitchy red; elytra testaceous, excepting their sutural, lateral and posterior margins; 3—4 lines; damp meadows. *Pl.* 21, *fig.* 4.

Genus [155]. CYPHA. *Kirby.* Six species.

    C. rufipes, *Kirb.* Black pubescent; antennæ and legs reddish; ⅔ of a line; common in moss. *Pl.* 21, *fig.* 5.

Genus [156]. CONURUS. *Steph.* Eleven species.

    C. pubescens, *Grav.* Pitchy black pubescent; apex and antennæ and legs testaceous; 2½ lines; meadows; not common. *Pl.* 21, *fig.* 6.

Genus [157]. TACHYPORUS. *Grav.* Thirty-four species.

    S. pyrrhopterus, *Kirb.* Black and shining; antennæ, margin of the thorax, elytra and legs reddish-testaceous; 1¼ line; meadows; not common. *Pl.* 21, *fig.* 7.

Genus [158]. BOLITOBIUS. *Leach.* Seventeen species.

    B. atricapillus, *Fab.* Reddish-testaceous, shining; head, middle of antennæ, elytra (excepting their shoulders and apical margin) and two terminal segments of abdomen, black; three lines; in moss and moist meadows. *Pl.* 21, *fig.* 8.

Genus [159]. MEGACRONUS. *Steph.* Eight species.

    M. analis, *Fab.* Black and shining; base and apex of antennæ, elytra, legs and apex of the abdomen testaceous; three lines; in moss and moist woods. *Pl.* 22, *fig.* 1.

Genus [160]. MYCETOPORUS. *Mann.* Eleven species.

    M. slpendens, *Marsh.* Deep black and very shining; antennæ fuscous with their base and apex testaceous; elytra, legs and apex of abdomen of a rich red; 2½—3 lines; humid meadows; common. *Pl.* 22, *fig.* 2.

Family [15]. ALEOCHARIDÆ. Shuck.

Genus [161]. DEINOPSIS. *Matth.* Six species.
> D. fuscatus, *Matth.* Pitchy black, opaque; tibiæ and tarsi fuscous; two lines; putrescent vegetables, rare. *Pl.* 22, *fig.* 3.

Genus [162]. CENTROGLOSSA. *Matth.* Six species.
> C. conuroides, *Matth.* Pitchy black opaque; legs fuscous; 1¼ line; putrescent vegetables. *Pl.* 22, *fig.* 4.

Genus [163]. MYLLÆNA. *Erich.* One species.
> M. dubia, *Grav.* Black opaque, covered with a silky cinereous pubescence; base of antennæ and legs yellowish; 1¼ line; rare. *Pl.* 22, *fig.* 5.

Genus [164]. DIGLOSSA. *Hal.* One species.
> D. mersa, *Hal.* Dull black pubescent; antennæ, palpi and legs fuscous; one line; sandy shores, on the coast of Ireland. *Pl.* 22, *fig.* 6.

Genus [165]. DINARDA. *Leach.* One species.
> D. dentata, *Grav.* Reddish chesnut, slightly shining; middle of the antennæ, head, disc of thorax, scutellum and base of the segments of the abdomen black; 1½ line; ant's nests; rare. *Pl.* 22, *fig.* 7.

Genus [166]. ATEMELES. *Dillwyn.* Two species.
> A. paradoxus, *Grav.* Opaque castaneous; head, excepting the mouth, black; 2¼ lines. Nests of *Formica rufa*, Lin. not uncommon. *Pl.* 22. *fig.* 8.

Genus [168]. ENCEPHALUS. *Kirb.* One species.
> E. complicans, *Kirb.* Black, shining; antennæ and legs pitchy; ¾ of a line; in moss. *Pl.* 23, *fig.* 1.

Genus [169]. GYROPHÆNA. *Mann.* Six species.
> G. nitidula, *Gyll.* Black, shining; base of the antennæ, legs and disc of the elytra testaceous; ¾ of a line; meadows. *Pl.* 23, *fig.* 2.

Genus [170]. OLIGATA. *Mann.* Six species.
> O. picipes, *Kirb.* Black, shining; legs pitchy; ¾ of a line; by sweeping in meadows and woods. *Pl.* 23, *fig.* 3.

Genus [171]. ALEOCHARA. *Grav.* Thirty-three species.
> A. fuscipes, *Payk.* Black, shining; elytra and legs chesnut red; three lines; meadows by sweeping. *Pl.* 23, *fig.* 4.

Genus [172]. OXYPODA. *Mann.* Sixteen species.
> O. lividipennis, *Mann.* Black, scarcely shining; elytra, apex of abdomen and legs fuscous; 1½ lines; meadows. *Pl.* 23. *fig.* 5.

Genus [173]. CALLICERUS. *Grav.* Two species?

C. Spencii, *Kirb.* Black; legs pitchy; head, thorax and elytra opaque; abdomen shining; 1¼ line; dead reeds, marshy places. *Pl.* 23, *fig.* 6.

Genus [174]. HOMALOTA. *Mann.* Twelve species?

H. plana, *Gyll.* Pitchy black; palpi, base of antennæ and legs fuscous; ¾ of a line; beneath bark; rare. *Pl.* 23, *fig.* 7.

Genus [175]. HYGRONOMA. *Erichs.* One species.

H. dimidiata, *Grav.* Black; base of antennæ, legs and posterior half of elytra yellow testaceous; 1¼ line; humid meadows by sweeping. *Pl.* 23, *fig.* 8.

Genus [176]. PHLÆOPORA. *Erichs.* Two species.

P. reptans, *Grav.* Black pubescent; antennæ at the base, apex of the abdomen and legs testaceous; elytra pitchy red; 1½ line; beneath bark; rare. *Pl.* 24, *fig.* 1.

Genus [177]. TACHYUSA. *Erichs.* One species.

T. atra, *Grav.* Black, opaque, covered with an ashy pubescence; legs pitchy; tarsi testaceous; 1¾ line; in moss; rare. *Pl.* 24, *fig.* 2.

Genus [178]. BOLITOCHARA. *Mann.* Seventy-eight species?

B. lunulata, *Payk.* Reddish chesnut; the disk of the elytra and the fifth and base of sixth segments of abdomen black; legs testaceous; two lines; sandy situations; not common. *Pl.* 24, *fig.* 3.

Genus [179]. OCALEA. *Erichs.* One species.

O. castanea, *Erichs.* Fusco-piceous; base of antennæ and legs testaceous; two lines; moss; rare. *Pl.* 24, *fig.* 4.

Genus [180]. CALODERA. *Mann.* Three species.

C. nigrita, *Mann.* Black, opaque; mouth and tarsi yellowish testaceous; two lines; humid places in woods; rare. *Pl.* 24, *fig.* 5.

Genus [181]. ZYRAS. *Steph.* One species.

Z. Haworthi, *Steph.* Red, shining; head, thorax, external angles of elytra and apex of abdomen black; 2½ lines; rare. *Pl.* 24, *fig.* 6.

Genus [182]. PELLA. *Steph.* Six species.

P. humeralis, *Grav.* Castaneous, shining; base of antennæ, humeral angles of elytra, apex of abdomen and legs paler; 2¼ lines; meadows by sweeping. *Pl.* 24, *fig.* 7.

Genus [183]. POLYSTOMA. *Steph.* One species.

P. obscurella, *Grav.* Pitchy black; head, thorax and elytra opaque; 1¾ lines; meadows. *Pl.* 24, *fig.* 8.

Genus [184]. ISCHNOPODA. *Steph.* Six species.

I. longitarsis, *Kirb.* Black, rather obscure; legs pitchy; 1¾ line; sandy coasts. *Pl.* 25, *fig.* 1.

Genus [185]. ASTILBUS. *Dillwyn.* One species.

A. canaliculatus, *Fab.* Castaneous; head, apex of antennæ and the third and fourth segments of the abdomen black; 2½ lines; sandy situations and moss. *Pl.* 25, *fig.* 2.

Genus [186]. FALAGRIA. *Leach.* Ten species.

F. nitens, *Kirb.* Pitchy black and shining; thorax sulcated; legs fuscous; one line; garden rubbish. *Pl.* 25, *fig.* 3.

Genus [187]. AUTALIA. *Leach.* Six species.

A. impressa, *Oliv.* Castaneous and shining; head, apex of antennæ, elytra and a ring near the apex of abdomen pitchy black; legs testaceous; thorax impressed behind; one line and a quarter; meadows by sweeping. *Pl.* 25, *fig.* 4.

Family [16]. PSELAPHIDÆ. Leach.

Genus [188]. CLAVIGER. *Preysler.* One species.

C. foveolatus, *Müll.* Entirely reddish, testaceous; one line; nests of Formica flava; rare. *Pl.* 25, *fig.* 5.

Genus [189]. EUPLECTUS. *Kirby.* Nine species.

E. nanus, *Reich.* Castaneous shining; legs red-testaceous; ¾ of a line; meadows by sweeping. *Pl.* 25, *fig.* 6.

Genus [190]. TRIMIUM. *Aubé.* One species.

T. brevicorne, *Reich.* Castaneous shining; ¾ of a line; meadows, roots of grass and moss. *Pl.* 25, *fig.* 7.

Genus [191]. BATRISUS. *Aubé.* One species.

B. venustus, *Reich.* Bright ferruginous, shining; abdomen black; 1½ line; moss, rare. *Pl.* 25, *fig.* 8.

Genus [192]. TYCHUS. *Leach.* One species.

T. niger, *Leach.* Black and shining; antennæ and legs pale pitchy; one line; moss in woods, common. *Pl.* 26, *fig.* 1.

Genus [193]. ARCOPAGUS. *Leach.* Four species.

A. glabricollis, *Reich.* Deep chesnut, shining; ¾ of a line; moss in meadows. *Pl.* 26, *fig.* 2.

Genus [194]. BYTHINUS. *Leach.* Four species.

B. securiger, *Reich.* Pitchy, shining; antennæ and legs bright ferruginous; ¾ of a line; moss and damp meadows. *Pl.* 26, *fig.* 3.

Genus [195]. BRYAXIS. *Kugel.* Seven species.

B. sanguineus, *Lin.* Black, antennæ fuscous; elytra blood red; legs pitchy brown; about one line; humid meadows. *Pl.* 26, *fig.* 4.

Genus [196]. PSELAPHUS. *Herbst.* Four species.
 P. Heiscii, *Herbst.* Chesnut brown and shining; antennæ and legs paler; 1¼ line; moss in winter. *Pl.* 26, *fig.* 3.

SUBDIVISION III. HELOCERA. SHUCK.

TRIBE I. CLAVICORNES. LATR.

Family [17]. SCYDMÆNIDÆ. Leach.

Genus [197]. SCYDMÆNUS. *Latr.* Seventeen species.
 S. tarsatus, *Kunz.* Pitchy brown; shining pubescent; antennæ and legs pale testaceous; 1¼ line; damp meadows by sweeping. *Pl.* 26, *fig.* 6.
Genus [198]. MEGALADERUS. *Stephens.* One species.
 M. thoracicus. Pitchy black, shining; antennæ and legs testaceous; length three-quarters of a line; moss in winter. *Pl.* 26, *fig.* 7.
Genus [199]. EUTHEIA. *Waterhouse.* One species.
 E. scydmænoides, *Waterh.* Black and shining; legs and antennæ testaceous; about half a line; meadows by sweeping and in moss. *Pl.* 26, *fig.* 8.

Family [18]. AGATHIDIIDÆ. West.

Genus [200]. SERICODERUS. *Steph.* One species.
 S. dubius, *Marsh.* Reddish; antennæ, thorax and legs testaceous; 1½ line; moist meadows. *Pl.* 27, *fig.* 1.
Genus [201]. ORTHOPERUS. *Steph.* Six species.
 O. punctum, *Marsh.* Obscure testaceous; head pitchy; antennæ and legs pale; about ½ a line; in garden rubbish. *Pl.* 27, *fig.* 2.
Genus [202]. CLYPEASTER. *Andersch.* One species.
 C. cassidoides, *Marsh.* Pitchy shining; margins of the thorax testaceous; antennæ and legs ferruginous; about ½ a line; beneath bark. *Pl.* 27, *fig.* 3.
Genus [203]. CLAMBUS. *Fisch.* Five species.
 C. armadillus, *De Geer.* Black and shining, antennæ and legs pale; about ½ a line. *Pl.* 27, *fig.* 4.
Genus [204]. AGATHIDIUM. *Illiger.* Fourteen species.
 A. seminulum, *Lin.* Pitchy black and shining; antennæ and legs reddish testaceous: about 1½ line. *Pl.* 27, *fig.* 5.
Genus [205]. LEIODES. *Latr.* Thirty species?

L. Cinnamomea, *Panz.* Reddish testaceous; antennæ and legs paler. In the female the posterior legs short and not curved; about 2½ lines; inhabits the truffle. *Pl.* 27, *fig.* 6.

Family [19]. SCAPHIDIIDÆ. Shuck.

Genus [206]. SCAPHIDIUM. *Oliv.* One species.
S. quadrimaculatum, *Oliv.* Black and shining, the elytra with four red spots; sometimes nearly obliterated; 2½ lines; not uncommon in fungi. *Pl.* 27, *fig.* 7.

Family [20]. CHOLEVIDÆ. Shuck.

Genus [207]. SCAPHISOMA. *Leach.* Two species.
S. agaricinum, *Lin.* Black and shining; antennæ and legs pale; apex of abdomen rufescent; nearly one line; fungi. *Pl.* 27, *fig.* 8.
Genus [208]. COLON. *Herbst.* One species.
C. brunneus, *Latr.* Deep cinnamon brown; antennæ and legs paler; very variable in colour; about a line; humid situations. *Pl.* 28, *fig.* 1.
Genus [209]. PTOMAPHAGUS. *Knoch.* Seven species?
P. truncatus, *Illiger.* Pitchy black; base of antennæ, tibiæ and tarsi reddish; about one line; humid meadows. *Pl.* 28, *fig.* 2.
Genus [210]. CATOPS. *Payk.* Seventeen species.
C. formicatus, *De Geer.* Blackish brown; legs pitchy ferruginous; about two lines; common, moist meadows. *Pl.* 28, *fig.* 3.
Genus [211]. CHOLEVA. *Latr.* Three species.
C. angustata, *Fab.* Pitchy brown; antennæ and legs a little paler; very variable in colour; two lines and a half; gardens and meadows. *Pl.* 28, *fig.* 4.

Family [21]. SPHÆRITIDÆ. Shuck.

Genus [212]. SPHÆRITES. *Duft.* One species.
C. glabratus. Black and shining, having above an æneous tinge; legs pitchy; two lines; Scotland, rare. *Pl.* 28, *fig.* 5.

Family [22]. NECROPHORIDÆ. Shuck.

Genus [213]. NECROPHORUS. *Fab.* Seven species.
N. ruspator, *Erichs.* (vestigator, *Steph.*) Black; the club
of the antennæ and two bands across the elytra (the last
of which is interrupted) of a bright orange red; ten lines;
carrion. *Pl.* 28, *fig.* 6.

Family [23]. SILPHIDÆ. Macleay.

Genus [214]. NECRODES. *Wilkin.* One species. In the female the thighs are simple and they are variable in size in the
male.
N. littoralis, *Lin.* Deep black; terminal joints of the antennæ orange and tarsi pitchy; 8—12 lines; carrion,
common. *Pl.* 28, *fig.* 7.
Genus [215]. OICEOPTOMA. *Leach.* Five species.
O. thoracica, *Lin.* Deep black; thorax of a deep orange;
seven lines; carrion, common. *Pl.* 28, *fig.* 1.
Genus [216]. SILPHA. *Linn.* Seven species.
S. nigrita, *Creutz.* Deep black, somewhat shining; six
lines; north of England. *Pl.* 29, *fig.* 1.
Genus [217]. PHOSPHUGA. *Leach.* Two species.
P. atrata, *Lin.* Deep black; five lines; common in fields
and pathways. *Pl.* 29, *fig.* 2.

Family [24]. NITIDULIDÆ. Macleay.

Genus [218]. THYMALUS. *Latr.* One species.
T. limbatus, *Fab.* Ferruginous, with a brassy reflection;
antennæ and legs rather paler; three lines; flowers in
spring and beneath bark. *Pl.* 29, *fig.* 3.
Genus [219]. NITIDULA. *Fab.* Twenty-six species.
N. grisea. *Lin.* Reddish testaceous, marbled with black;
2½—3 lines; common, under bark. *Pl.* 29, *fig.* 4.
Genus [220]. CRYPTARCHA. *Shuck.* Two species.
C. strigata, *Fab.* Pitchy black with pale markings on the
elytra; antennæ and legs rufo-testaceous; beneath bark.
Pl. 29, *fig.* 5.
Genus [221]. STRONGYLUS. *Herbst.* Two species.
T. ferruginous, *Fab.* Ferruginous red; variable, being
either paler or darker; about 2½ lines; common in fungi.
Pl. 29, *fig.* 6.

25

Genus [222]. CAMPTA. *Kirby.* One species.
    C. lutea, *Herbst.* Yellow testaceous; club of the antennæ
darker; 2½ lines; common in flowers. *Pl.* 29, *fig.* 7.
Genus [223]. MELIGETHES. *Kirby.* Eleven species.
    M. rufipes, *Dej.* Black; the legs rufo-testaceous; about two
lines; common in flowers. *Pl.* 29, *fig.* 8.
Genus [224]. PRIA. *Kirby.* One species.
    P. truncatella, *Marsh.* Entirely fuscous; ¾ of a line; mid-
summer, in flowers. *Pl.* 30, *fig.* 1.
Genus [225]. ANOMÆOCERA. *Shuck.* [Anisocera, *Howit.*] One
species. The female has the second joint of the antennæ less
developed.
    A. Spiræ, *Howit.* Reddish testaceous; the disk of the
elytra sometimes darker; about one line; the North on the
Spiræa Ulmata. *Pl.* 30, *fig.* 2.
Genus [226]. CATERETES. *Herbst.* Twelve species.
    C. bipustulatus, *Pkl.* Pitchy black; antennæ, legs and a
spot on each of the elytra testaceous; about one line; by
sweeping in marshy meadows. *Pl.* 30, *fig.* 3.
Genus [227]. TRICHOPTERYX. *Kirby.* Seven species.
    T. atomaria, *De Geer.* Pitchy black pubescent; antennæ
and legs testaceous; about ½ a line; rotting vegetables
and hot beds. *Pl.* 30, *fig.* 4.
Genus [228]. MICROPEPLUS. *Latr.* Three species?
    M. staphylinoïdes, *Marsh.* Black; antennæ, sides of the
thorax, and legs ferruginous; nearly a line; meadows
by sweeping. *Pl.* 30, *fig.* 5.
Genus [229]. CARPOPHILUS. *Leach.* Two species.
    C. hemipterus, *Lin.* Pitchy black; a spot on the shoulder
and another larger towards the apex of the elytra yellow
testaceous; legs testaceous; rather more than a line; a
doubtful native. *Pl.* 30, *fig.* 6.

Family [25]. ENGIDÆ. Macleay.

Genus [230]. IPS. *Fab.* Four species.
    I. quadripustulatus, *Linn.* Depressed; deep black and
shining; each elytra with two ferruginous spots, one
humeral and the other placed about the middle; 3½ lines;
under the bark of the pine; Scotland, rare. *Pl.* 30,
fig. 7.
Genus [230a]. PITYOPHAGUS. *Shuck.* One species.
    P. ferruginous, *Lin.* Cylindrical and entirely of a ferrugi-
nous red; the head a little darker; 2—2½ lines; beneath
bark, rare. *Pl.* 30, *fig.* 8.

E

Genus [231]. CRYPTOPHAGUS. *Herbst.* Fourteen species.

 C. fumatus, *Marsh.* Entirely reddish testaceous, covered with a short pubescence ; one line ; by sweeping, and flying towards sunset. *Pl.* 30, *fig.* 9.

Genus [232]. ANTHEROPHAGUS. *Knoch.* Three species.

 A. pallens, *Lin.* Pale yellow testaceous ; antennæ, base of the tibiæ and tarsi pitchy ; 2—2½ lines ; flowers in June. *Pl.* 31, *fig.* 1.

Genus [233]. ANISARTHRIA. *Waterhouse.* Eight species ?

 A. nitida, *Steph.* Bright shining black ; antennæ and legs testaceous ; one third of a line ; garden rubbish, and flying in its vicinity on warm evenings. *Pl.* 31, *fig.* 2.

Genus [234]. ATOMARIA. *Kirb.* Twenty species ?

 A. nigripennis, *Payk.* Black ; the head, thorax, antennæ and legs ferruginous ; two thirds of a line ; meadows by sweeping. *Pl.* 31, *fig.* 3.

Genus [235]. ENGIS. *Paykul.* Three species.

 E. scanicus, *Lin.* Black, brightly shining ; head, thorax, humeral angles of elytra, antennæ and legs bright ferruginous ; 1½ line ; fungi, not common. *Pl.* 31, *fig.* 4.

Genus [236]. TYPHÆA. *Kirby.* Three species.

 T. ferruginea, *Marsh.* Pubescent and entirely ferruginous ; rather more than a line ; beneath bark. *Pl.* 31, *fig.* 5.

Genus [237]. MYCETÆA. *Kirby.* Two species.

 M. fumata, *Marsh.* Pale testaceous, slightly pubescent ; three quarters of a line ; by sweeping. *Pl.* 31, *fig.* 6.

Genus [238]. PARAMECOSOMA. *Curt.* One species.

 T. bicolor, *Curt.* Fusco-ferruginous ; head and thorax black ; three quarters of a line ; north of England. *Pl.* 31, *fig.* 7.

Genus [239]. CORTICARIA. *Marsh.* Eleven species ?

 C. elongata, *Illig.* Pale ferruginous ; pubescent ; about a line ; by sweeping. *Pl.* 31, *fig.* 8.

Genus [240]. HOLOPARAMECUS. *Curt.* One species.

 H. depressus, *Curt.* Bright shining testaceous ; one third of a line : a doubtful native. *Pl.* 32, *fig.* 1.

Genus [241]. TETRATOMA. *Herbst.* Five species.

 T. Ancora, *Fab.* Pale testaceous, shining ; the elytra having a cordate spot near the scutellum, and lateral markings, blackish ; 1½ line ; beneath bark ; rare. *Pl.* 32, *fig.* 2.

Genus [242]. MYCETOPHAGUS. *Fab.* Six species.

 M. 4-pustulatus, *Lin.* Pitchy black, pubescent ; legs and four spots on the elytra ferruginous ; three lines and a quarter ; abundant in putrescent fungi. *Pl.* 32, *fig.* 3.

Genus [243]. BIPHYLLUS. *De Jean.* One species.

27

B. lunatus, *Fab.* Pitchy black pubescent ; a common lunate spot about the middle of the elytra, produced by decumbent silvery pile ; antennæ and legs ferruginous ; 1½ line ; beneath bark, rare. *Pl. 32, fig.* 4

Genus [244]. TRIPHYLLUS. *Megerle.* Two species.
T. punctatus, *Fab.* Pitchy black ; head, thorax, base and apex of elytra, antennæ and legs ferruginous ; 1½ line ; fungi. *Pl. 32, fig.* 5.

Genus [245]. PHLOIOPHILUS. *Waterhouse.* Three species?
P. Cooperi, *Steph.* Pale testaceous, with dark markings on the elytra ; one line ; beneath bark ; rare. *Pl. 32, fig.* 6.

Genus [246]. BYTURUS. *Latr.* One species.
B. tomentosus, *Fab.* Testaceous or fuscous ; covered with a close decumbent yellow pile ; two lines ; common in flowers of the bramble in June. *Pl. 32, fig.* 7.

Genus [247]. CERYLON. *Latr.* Three species?
C. histeroides, *Panz.* Pitchy black and shining ; head, thorax, antennæ and legs deep ferruginous ; one line ; beneath bark. *Pl. 32, fig.* 8.

Genus [248]. SYNCHITA. *Helwig.* One species.
S. Juglandis, *Fab.* Pitchy ferruginous ; antennæ slightly paler ; 1¾ line ; beneath the bark of the walnut, rare. *Pl. 33, fig.* 1.

Genus [249]. ANOMMATUS. *Wesmael.* One species.
A. obsoletum, *Spence.* Testaceous or castaneous ; very shining and convex ; three quarters of a line ; beneath stones. *Pl. 33, fig.* 2.

Genus [250]. RHYZOPHAGUS. *Herbst.* Ten species.
R. ferruginous, *Payk.* Entirely deep, or pale ferruginous and shining ; about two lines ; beneath bark. *Pl. 33, fig.* 3.

Genus [251]. LISSODEMA. *Curt.* One species.
L. Heyana, *Curt.* Pitchy chesnut, shining, antennæ and legs slightly paler ; one line and one third ; Derbyshire, rare. *Pl. 33, fig.* 4.

Genus [252]. MONOTOMA. *Herbst.* Three species?
M. picipes, *Herbst.* Pitchy black, or testaceous ; antennæ and legs reddish testaceous ; about one line ; meadows by sweeping, and garden rubbish. *Pl. 33, fig.* 5.

Genus [253]. CICONES. *Curt.* One species.
C. variegata, *Helwig.* Pitchy black, pubescent ; elytra variegated with ferruginous spots ; antennæ and legs also ferruginous ; 1½ line ; beneath bark, rare. *Pl. 33, fig.* 6.

Genus [254]. BITOMA. *Herbst.* One species.
B. crenata. *Fab.* Deep black ; elytra with four blood red spots, two placed at the humeral angles and two at the

apex; antennæ and legs ferruginous; 1½ line; beneath bark. *Pl.* 33, *fig.* 7.

Genus [255]. LATRIDIUS. *Herbst.* Eleven species?

L. lardarius, *De Geer.* Red testaceous, head and thorax blood red; one line; garden rubbish. *Pl.* 33, *fig.* 8.

Genus [256]. SILVANUS. *Latr.* One species.

S. unidentatus, *Fab.* Red testaceous; antennæ and legs slightly paler; 1½ line; beneath bark; rare. *Pl.* 34, *fig.* 1.

Genus [257]. PEDIACUS. *Shuck.* One species.

P. dermestoides, *Fab.* Very flat; castaneous brown; the head and thorax rather the darkest; two lines; beneath bark, rare. *Pl.* 34, *fig.* 2.

Genus [258]. TROGOSITA. *Oliv.* One species.

T. Mauritanica, *Lin.* Pitchy black and shining; 3½—4 lines; in houses, rarely at large. *Pl.* 34, *fig.* 3.

Genus [259]. NEMOSOMA. *Desmarests.* One species.

N. elongata, *Lin.* Cylindrical; pitchy black; base of the elytra and an obsolete spot towards their apex and the antennæ and legs testaceous; two lines; beneath bark, rare. *Pl.* 34, *fig.* 4.

Genus [260]. COLYDIUM. *Herbst.* One species.

C. elongatum, *Fab.* Shining black; antennæ and legs pitchy red; nearly four lines; beneath bark; New Forest, rare. *Pl.* 34, *fig.* 5.

Genus [260a]. TEREDUS. *Dejean.* One species.

T. nitidus, *Helwig.* Deep pitchy black and very shining; antennæ and legs ferruginous; rather more than two lines; beneath bark, Sherwood Forest, rare. *Pl.* 34, *fig.* 6.

Genus [261]. XYLOTROGUS. *Steph.* One species.

X. brunneus, *Steph.* Entirely brown; antennæ and legs ferruginous; 2½ lines; rare. *Pl.* 34, *fig.* 7.

Genus [262]. LYCTUS. *Fab.* One species.

L. canaliculatus, *Fab.* Entirely chesnut brown; 2½ lines; oak palings; abundant in June. *Pl.* 34, *fig.* 8.

Family [26]. DERMESTIDÆ. *Leach.*

Genus [263]. THROSCUS. *Latr.* Two species?

T. dermestoides, *Lin.* Dark chesnut, pubescent; 1½—2 lines; palings in woods. *Pl.* 35, *fig.* 1.

Genus [264]. DERMESTES. *Lin.* Five species.
    D. murinus, *Lin.* Raven black, mottled with ashy hairs; three and a half lines; dead animals; common. *Pl.* 35, *fig.* 2.
Genus [265]. TIRESIAS. *Step.* One species.
    T. serra, *Fab.* Deep black and shining, slightly pubescent; antennæ and legs ferruginous; two lines; beneath bark. *Pl.* 35, *fig.* 3.
Genus [266]. ATTAGENUS. *Latr.* Two species.
    A. pellio, *Lin.* Pitchy black, pubescent, with a slight ashy patch of hair at the three posterior angles of the thorax, and another rather larger on the disc, towards the middle of each elytron; one and a half to two and a half lines; in houses. *Pl.* 35, *fig.* 4.
Genus [267]. MEGATOMA. *Herbst.* One species.
    M. undata, *Lin.* Raven black, with a small patch of silvery grey hair at the posterior angles of the thorax, and two transverse crenulated bands of the same colour across the elytra; two to two and a half lines; palings; not uncommon. *Pl.* 35, *fig.* 5.

Family [27]. BYRRHIDÆ. Leach.

Genus [268]. ASPIDIPHORUS. *Ziegler.* One species.
    A. orbiculatus, *Gyll.* Black and shining; legs reddish testaceous; rather more than half a line; widely dispersed, but rare; in moss. *Pl.* 35, *fig.* 6.
Genus [269]. ANTHRENUS. *Geoffroy.* Five species?
    A. Scrophulariæ, *Lin.* Black, with the posterior angles of the thorax covered with whitish pubescence, and three transverse irregular bands of the same colour on the elytra, which near the suture and the sutural markings bright red; tibiæ and tarsi ferruginous; one line and a half; rare. *Pl.* 35, *fig.* 7.
Genus [270]. TRINODES. *Megerle.* One species.
    Tr. hirtus, *Fab.* Black, shining, covered with tolerably long pubescence; legs and antennæ testaceous; one line and a half; beneath bark; Notts, Windsor, and Exeter. *Pl.* 36, *fig.* 1.
Genus [271]. LIMNICHUS. *Ziegler.* One species.
    S. sericeus, *Duffts.* Black, covered with a silky grey pubescence; legs and antennæ piceous; nearly one line; grassy places on the coasts by sweeping. *Pl.* 36, *fig.* 2.
Genus [272]. SYNCALYPTA. *Dillwyn.* Four species.
    S. arenaria, *Sturm.* Black, slightly shining; covered with

dispersed erect, rigid, capitate setæ ; nearly a line ; in sandy and chalky situations. *Pl.* 36, *fig.* 3.

Genus [273]. NOSODENDRON. *Latr.* One species.

N. fasciculare, *Oliv.* Black and shining ; elytra covered with fascicles of reddish brown hair placed in rows ; antennæ and legs pitchy ; nearly three lines ; Southend, beneath the bark of elms. *Pl.* 36, *fig.* 4.

Genus [274]. BYRRHUS. *Lin.* Nine species.

B. pilula, *Lin.* Pitchy black, covered with a dense decumbent silky down of a bright brown, with alternate darker stripes variously interrupted ; antennæ, legs and underside varying from black to red ; extremely variable in markings ; four to five lines ; sandy situations. *Pl.* 36, *fig.* 5.

Genus [275]. OOMORPHUS. *Curtis.* One species.

O. concolor, *Sturm.* Deep black and shining ; one line and a half ; sandy and chalky places on the coasts. *Pl.* 36, *fig.* 6.

Genus [276]. SIMPLOCARIA. *Marsh.* Two species.

S. semistriata, *Illig.* Bronzy black, shining, covered with a silky pubescence ; antennæ and legs testaceous ; about a line and a half ; sandy and grassy places. *Pl.* 36, *fig.* 7.

Genus [277]. EPHISTEMUS. *West.* Four species ?

E. gyrinoides, *Marsh.* Black, smooth and shining ; apex of elytra reddish ; head in front, antennæ and legs testaceous ; half a line ; grassy places by sweeping. *Pl.* 36, *fig.* 8.

Family [28]. HETEROCERIDÆ. Macleay.

Genus [278]. HETEROCERUS. *Bosc.* Seven species ?

H. marginatus, *Fab.* Dusky black ; very pubescent ; elytra with the margin and several obscure red spots (frequently wanting) on the disk ; variable ; antennæ and tarsi pale or obscure red ; three lines ; margin of ponds and ditches. *Pl.* 37, *fig.* 1.

Family [29]. PARNIDÆ. Macleay.

Genus [279]. PARNUS. *Fab.* Four species ?

P. prolefericornis, *Fab.* Dusky black, covered with a short dense pubescence ; legs pitchy ; two and a half lines ; banks of ponds. *Pl.* 37, *fig.* 2.

31

Genus [280]. DRYOPS. *Oliv.* One species.
    D. Dumerilii, *Latr.* Dusky black; covered with a short
close pubescence; antennæ and tarsi reddish; two and
three quarters lines; banks of the Waudle, Surrey; and
banks of the Trent, Notts. *Pl.* 37, *fig.* 3.

Family [30]. ELMIDÆ. Shuck.

Genus [281]. ELMIS. *Latr.* Eleven species.
    E. Volkmari, *Panz.* Entirely bronzy black; one and a
half line; beneath stones in rapid streams. *Pl.* 37, *fig.* 4.
Genus [282]. GEORYSSUS. *Latr.* One species.
    G. pygmæus, *Fab.* Deep black and shining; head de-
flexed; the punctures of the striæ of the elytra very
coarse; three quarters of a line; muddy banks of oozing
springs. *Pl.* 37, *fig.* 5.

TRIBE II. PALPICORNES. LATR.

Family [31]. SPERCHEIDÆ. Shuck.

Genus [283]. SPERCHEUS. *Fab.* One species.
    S. emarginatus, *Fab.* Obscure testaceous; opaque; elytra
with dispersed dusky spots; antennæ and legs testaceous;
three and a half lines; roots of aquatic plants in stagnant
waters. *Pl.* 37, *fig.* 6.

Family [32]. HELOPHORIDÆ. Leach.

Genus [284]. HELOPHORUS. *Illig.* Eleven species?
    H. grandis, *Illig.* Griseous, loosely sprinkled with dusky
spots; the channels of the thorax cupreous; antennæ,
palpi and legs testaceous; about three lines; in ponds,
very common. *Pl.* 38, *fig.* 1.
Genus [285]. HYDROCHUS. *Germ.* Three species.
    H. elongatus, *Fab.* Entirely of a shining bronzy black;
antennæ and legs pitchy; two lines; ponds and ditches.
Pl. 38, *fig.* 2.
Genus [286]. ENICOCERUS. *Stephens.* Three species?
    E. viridiæneus, *Steph.* Greenish brassy, shining; antennæ,
palpi and legs pitchy; about a line and a half; under
stones in streams. *Pl.* 38, *fig.* 3.

Genus [287]. OCHTHEBIUS. *Leach.* Ten species.
>O. marinus, *Payk.* Brassy green and shining; antennæ and legs testaceous; about a line; stagnant waters. *Pl.* 38, *fig.* 4.

Genus [288]. AMPHIBOLUS. *Waterh.* One species.
>A. atricapillus, *Waterh.* Testaceous, with a coppery reflection; head black; antennæ, palpi and legs dusky testaceous; about a line long; running streams, Yorks. *Pl.* 38, *fig.* 5.

Genus [289]. HYDRÆNA. *Kugelann.*
>H. riparia, *Kug.* Pitchy black, shining; antennæ, palpi and legs reddish testaceous; one line and a quarter; running streams. *Pl.* 38, *fig.* 6.

Family [33]. HYDROPHILIDÆ. Leach.

Genus [290]. LIMNEBIUS. *Leach.* Nine species.
>L. truncatellus, *Fab.* Deep black; antennæ, palpi and legs reddish; one line and a quarter; ponds, ditches, and streams. *Pl.* 39, *fig.* 1.

Genus [291]. LACCOBIUS. *Erichson.* One species.
>L. minutus, *Lin.* Black; head and thorax brassy; elytra dirty testaceous, sprinkled with dark spots; antennæ and legs pale; one line and a half; ponds and ditches. *Pl.* 39, *fig.* 2.

Genus [292]. BEROSUS. *Leach.* Four species.
>B. luridus, *Lin.* Obscure testaceous; head, and a square spot on the thorax brassy; elytra with dispersed dark spots; legs pale, their tarsi darker; two and a half to three lines; ponds and ditches. *Pl.* 39, *fig.* 3.

Genus [293]. HYDROÜS. *Lin.* One species. The tarsi of the female are simple.
>H. piceus, *Lin.* Entirely of a greenish black, shining; antennæ reddish pitchy; legs pitchy; one inch and three to six lines; stagnant ponds and ditches. *Pl.* 39, *fig.* 4.

Genus [294]. HYDROPHILUS. *Fab.* One species.
>H. caraboides, *Lin.* Greenish black and shining; antennæ and palpi pitchy red; legs pitchy; about ten lines; ponds and ditches. *Pl.* 39, *fig.* 5.

Genus [295]. HYDROBIUS. *Leach.* Five species.
>H. oblongus, *Herb.* Pitchy black and shining; antennæ, palpi, tibiæ and tarsi reddish; four lines; bank of ponds. *Pl.* 40, *fig.* 1.

Genus [296]. PHILHYDRUS. *Solier.* Fifteen species.
>P. melanocephalus, *Fab.* Testaceous; head, disk of thorax,

and two small spots at its margins, and elytra, with the
shoulders, and a sutural stripe, all black ; antennæ and
legs reddish ; two and a half lines; ponds and ditches.
Pl. 40, *fig.* 2.

Genus [297]. CHÆTARTHRIA. *Waterh.* One species.

Ch. seminulum, *Payk.* Black and shining ; antennæ, apex
of elytra and legs pitchy red ; about a line ; ponds and
ditches. *Pl.* 40, *fig.* 3.

Family [34]. SPHÆRIDIIDÆ. Leach.

Genus [298]. CYCLONOTUM. *Erichs.* One species.

C. orbiculare, *Fab.* Deep black and shining ; antennæ
and legs pitchy ; about two lines; ponds and ditches.
Pl. 40, *fig.* 4.

Genus [299]. SPHÆRIDIUM. *Fab.* Three species.

S. scarabæoides, *Lin.* Black and shining ; elytra with a
red humeral spot and their apex ochraceous; antennæ
and legs pitchy red ; two to three and a half lines ; dung
of cows and horses. *Pl.* 40, *fig.* 5.

Genus [300]. CERCYON. *Leach.* Fifty-five species?

C. obsoletum, *Gyll.* Black and shining; apex of elytra,
antennæ, palpi and legs pitchy red; one line and three-
quarters ; dung of horses. *Pl.* 40, *fig.* 6.

TRIBE III. FRACTICORNES. *Shuck.*

Family [35]. HISTERIDÆ. Leach.

Genus [301]. PLATYSOMA. *Leach.* Three species.

P. depressum, *Fab.* Deep black and shining; antennæ
and legs pitchy red; one line and a half ; under bark.
Pl. 41, *fig.* 1.

Genus [302]. HISTER. *Linn.* Twenty-four species?

H. unicolor, *Lin.* Entirely deep black and shining ; from
two and a half to four lines; in the dung of animals.
Pl. 41, *fig.* 2.

Genus [303]. DENDROPHILUS. *Leach.* Six species.

D. quatuordecim striatus, *Steph.* Deep black and shining;
antennæ and legs pitchy; one line and a half ; in rubbish
heaps, garden refuse, &c. *Pl.* 41, *fig.* 3.

[This species, which is here figured as a *Dendrophilus*,
is an *Epierus* of Erichson: a true Dendrophilus will
be figured in the Supplement.]

F

Genus [304]. PAROMALUS. *Erichs.* Two species.
    P. flavicornis, *Herbst.* Deep black and shining; antennæ
    testaceous; legs pitchy red; about one line; in boleti
    and beneath bark. *Pl.* 41, *fig.* 4.
Genus [305]. SAPRINUS. *Erichs.* Eleven species.
    S. nitidulus, *Fab.* Bronzy black, very shining; legs pitchy;
    two lines and a half; dung of animals. *Pl.* 41, *fig.* 5.
Genus [305ª]. TERETRIUS. *Erichs.* One species.
    T. picipes, *Fab.* Deep black and very shining; antennæ
    and legs pitchy red; about a line; oak palings, beneath
    the bark; Camberwell, Hampstead. *Pl.* 41, *fig.* 6.
Genus [306]. ONTHOPHILUS. *Leach.* Two species.
    O. sulcatus, *Fab.* Deep opaque black; one and a half to
    two lines; dung of animals; Nottinghamshire, Coombe
    Wood. *Pl.* 41, *fig.* 7.
Genus [307]. ABRÆUS. *Leach.* Two species?
    A. globosus, *Payk.* Pitchy black and very shining; about
    one line and a half; decaying vegetables and garden
    refuse. *Pl.* 41. *fig.* 8.

SUBDIVISION IV. PETALOCERA. SHUCK.

TRIBE I. PECTENICORNES. SHUCK.

Family [36]. LUCANIDÆ. Leach.

Genus [308]. PLATYCERUS. *Geoffr.* One species.
    P. caraboides, *Lin.* Bright blue and shining; antennæ
    and legs black; six lines and a half; West of England.
    Pl. 42, *fig.* 1.
Genus [309]. DORCUS. *Macleay.* One species.
    D. parallelipipedus, *Lin.* Entirely of an opaque black;
    about twelve lines; woods, in rotten trees. *Pl.* 42,
    fig. 2.
Genus [310]. LUCANUS. *Lin.* One species. In the female
the mandibles are small.
    L. cervus, *Lin.* Black and slightly shining; mandibles and
    elytra usually of a dark chesnut; from one to two and a
    half inches; common in the South of England, in the
    vicinity of woods. *Pl.* 42, *fig.* 3.
Genus [311]. SINODENDRON. *Fab.* One species. In the
female the horn of the head is obsolete, and the thorax is less
retuse.
    S. cylindricum, *Lin.* Either dark black and shining, or
    castaneous; six to eight lines; common in rotten willows.
    Pl. 42, *fig.* 4.

Tribe II. LAMELLICORNES. Shuck.

Race I. SAPROPHAGA. Macleay.

Family [37]. Geotrupidæ. Leach.

Genus [312]. Geotrupes. *Latr.* Nine species?
G. stercorarius, *Lin.* Deep black; inside of the legs steel blue; nine to twelve lines; common in meadows. *Pl.* 43, *fig.* 1.
Genus [313]. Typhæus. *Leach.* One species. The thoracic spines in the female are obsolete.
T. vulgaris, *Leach.* Deep black, slightly shining; six to eight lines; common on sandy heaths. *Pl.* 43, *fig.* 2.
Genus [314]. Bolbocerus. *Kirby.* One species. The spines of the head and thorax are obsolete in the female.
B. mobilicornis, *Fab.* Deep black and shining, or testaceous; four lines; heaths and sandy districts, uncommon. *Pl.* 43, *fig.* 3.

Family [38]. Scarabæidæ. Macleay.

Genus [315]. Copris. *Geoffr.* One species. The horn of the head is obsolete and emarginate in the female.
C. lunaris, *Lin.* Bright shining black; ten lines; sandy heaths; not common. *Pl.* 43, *fig.* 4.
Genus [316]. Onthophagus. *Latr.* Nine species. The usual horn of the head is obsolete in the female.
O. nuchicornis, *Lin.* Bronzy black; elytra testaceous, mottled with bronzy spots; about three lines; in dung, especially in sandy situations. *Pl.* 43, *fig.* 5.

Family [39]. Aphodiidæ. Macleay.

Genus [317]. Aphodius. *Illig.* Fifty-eight species?
A. fossor, *Lin.* Entirely black and shining; four to six lines; common in dung. *Pl.* 43, *fig.* 6.
Genus [318]. Oxyomous. *Esch.* Differs from Aphodius only in the trophi, and is therefore not figured.
Genus [319]. Psammodius. *Gyllenhal.* Two species.
P. sulcicollis, *Illig.* Black and shining, or castaneous; legs pitchy red; about two lines; sandy places, especially the coasts. *Pl.* 43, *fig.* 7.

Family [40]. TROGIDÆ. Macleay.

Genus [320]. ÆGIALIA. *Latr.* One species.
    Æ. globosa, *Illig.* Pitchy black or castaneous; from two to two and a half lines; sandy places, especially the coasts. *Pl.* 44, *fig.* 1.

Genus [321]. TROX. *Fab.* Four species.
    T. sabulosus, *Lin.* Dull opaque black; antennæ and legs pitchy; about four lines; sandy and gravelly heaths. *Pl.* 44, *fig.* 2.

RACE II. THALEROPHAGA. MACLEAY.

Family [41]. MELOLONTHIDÆ. Macleay.

Genus [322]. SERICA. *Macleay.* One species. The female has the club of the antennæ short.
    S. brunnea, *Lin.* Reddish testaceous, with a slight opaline tinge; vertex pitchy; about five lines; sandy situations. *Pl.* 44, *fig.* 3.

Genus [323]. OMALOPLIA. *Megerle.* One species.
    O. ruricola, *Fab.* Deep black; elytra reddish testaceous, excepting their suture and margins, which are black; legs pitchy red; about four lines; hedges, near woods. *Pl.* 44, *fig.* 4.

Genus [324]. RHISOTROGUS. *Latr.* One species. Club of the antennæ in the female short.
    R. solstitialis, *Lin.* Pale testaceous, very pubescent; antennæ and legs reddish testaceous; about nine lines; elms and hedges; common. *Pl.* 44, *fig.* 5.

Genus [325]. MELOLONTHA. *Fab.* Three species. Club of the antennæ in the female short.
    M. vulgaris, *Lin.* Red testaceous, pubescent; head, thorax, and scutellum black; ten to twelve lines; hedges in fields; very common. *Pl.* 44, *fig.* 6.

Genus [326]. PHYLLOPERTHA. *Kirby.* Three species.
    P. horticola. Very pubescent; head, thorax, scutellum and legs bright shining green; elytra testaceous, with their suture and margins narrowly edged with black; four to five lines; hedges in fields; common. *Pl.* 45, *fig.* 1.

Genus [327]. ANOMALA. *Megerle.* Two species.
    A. Frischii, *Fab.* Bright shining green, coppery green, or blue, with the lateral margins of the thorax and the elytra testaceous; legs usually blue, black, or coppery. Sometimes entirely green, coppery, or blue; six and a

half to seven and a half lines ; sandy places, especially the coasts. *Pl.* 45, *fig.* 2.

Genus [328]. ANISOPLIA. *Megerle.* One species.

    A. Agricola, *Lin.* Bronzy ; elytra testaceous ; the margins, a transverse band in the middle, and the suture (broadest between the scutellum) black, as well as the antennæ and legs ; seven lines ; South Wales. *Pl.* 45. *fig.* 3.

Genus [329]. HOPLIA. *Illig.* One species. The posterior legs are shorter in the female.

    H. argentea, *Oliv.* Brownish black ; the female with the base of the antennæ, elytra and legs red ; male with the legs black, excepting their tarsi which are pitchy red ; antennæ also of this colour ; three to four and a half lines ; sandy heaths, common. *Pl.* 45, *fig.* 4.

Family [42]. CETONIIDÆ. Macleay.

Genus [330]. TRICHIUS. *Fab.* One species.

    T. fasciatus, *Lin.* Densely covered, excepting the elytra, with long fulvous hair ; head, thorax (excepting the sides of the latter behind, which are yellow), scutellum and legs black ; elytra reddish testaceous, with six irregular large spots placed externally and the suture black ; about seven lines ; South Wales. *Pl.* 45, *fig.* 5.

Genus [331]. GNORIMUS. *St. Farg.* Two species. The intermediate tibiæ in the female are neither clavate nor arcuate.

    G. nobilis, *Lin.* Entirely of a rich shining golden or coppery green, with usually several white spots upon the elytra and pygidium ; eight to ten lines ; rotten apple trees and flowers, especially those of the alder. *Pl.* 45, *fig.* 6.

Genus [332]. CETONIA. *Fab.* Two species ?

    C. aurata, *Lin.* Entirely of a rich shining golden or coppery green, varied beyond the centre of the elytra with abbreviated transverse and slightly waved white lines ; eight to eleven lines ; very common in gardens. *Pl.* 45, *fig.* 7.

SUBDIVISION V. PRIONOCERA. SHUCK.

TRIBE I. STERNOXI. LATR.

Family [43]. BUPRESTIDÆ. Leach.

Genus [333]. ANTHAXIA. *Eschholtz.* Two species.

    A. Salicis, *Fab.* Head, thorax and base of elytra in a

semicircle from shoulder to shoulder of a rich metallic blue or green; the remainder of the elytra of a bright golden red, varying into purple ; antennæ black ; legs coloured like the thorax ; about three lines; willows, very rare. *Pl.* 46, *fig.* 1.

Genus [334]. AGRILUS. *Megerle.* Five species.

A. biguttatus, *Lin.* Of a rich shining metallic blue or green, with a white spot on each elytron near the suture, and another on the lateral projecting portions of the first dorsal segment of the abdomen; five to six and a half lines; Darenth and Hampstead, not common. *Pl.* 46, *fig.* 2.

Genus [335]. APHANISTICUS. *Latr.* One species.

A. pusillus, *Oliv.* Entirely bronzy black ; nearly two lines; grassy and mossy places in woods. *Pl.* 46, *fig.* 3.

Genus [336]. TRACHYS. *Fab.* Three species.

T. minuta, *Lin.* Bronzy or brassy, with several whitish waved markings and transverse bands, especially towards the apex of the elytra ; nearly two lines; foliage in woods. *Pl.* 46, *fig.* 4.

Family [44]. MELASIDÆ. Shuck.

Genus [337]. MELASIS. *Oliv.* One species. In the male the antennæ are more strongly pectinated and the anterior angles of the thorax rounded.

M. buprestoides, *Lin.* Opaque, brown, black, or cinnamon brown ; antennæ and legs a little paler ; two and a half to four and a half lines ; Sherwood, Windsor, and New Forests ; rotten stumps and trees. *Pl.* 46, *fig.* 5.

Family [45]. EUCNEMIDÆ. Westw.

Genus [338]. MICRORHAGUS. *Eschholtz.* One species.

M. pygmæus, *Fab.* Entirely of a shining brownish black ; about two and a quarter lines ; Norfolk? *Pl.* 46, *fig.* 6.

Family [46]. ELATERIDÆ. Leach.

Genus [339]. ADRASTUS. *Esch.* Two species.

A. limbatus, *Fab.* Head and thorax black and shining ; elytra dark dirty testaceous, except the lateral margins and suture, which are pitchy, as well as the antennæ and legs ; two and a quarter lines ; meadows and pastures, common. *Pl.* 46, *fig.* 7.

Genus [340]. DOLOPIUS. *Esch*. One species.

D. marginatus, *Lin*. Head black; thorax brown on the disk, the margins brownish testaceous; elytra dark testaceous, with the suture broadly brown, narrowing towards the apex; antennæ and legs testaceous; three to four lines; meadows and grassy places in woods. *Pl. 47, fig.* 1.

Genus [341]. AGRIOTES. *Esch*. Four species.

A. lineatus, *Lin*. Entirely of an obscure fuscous; thorax and the alternate striæ of the elytra rather darker; about four and a half lines; meadows and pastures, common. *Pl. 46, fig.* 8.

Genus [342]. SERICOSOMUS. *Serville*. Four species.

S. brunneus, *Lin*. Cinnamon brown, with the head, antennæ, disk of thorax and scutellum blackish brown; about five lines; grassy places in woods, not common. *Pl. 47, fig.* 2.

Genus [343]. ECTINUS. *Esch*. One species.

E. aterrimus, *Lin*. Intensely black; legs sometimes pitchy; about six lines; grassy places, rare. *Pl. 47, fig.* 3.

Genus [344]. LIMONIUS. *Esch*. Five species.

L. serraticornis, *Payk*. Black, with a slight metallic tinge; legs obscure testaceous; about three lines; not common; by beating in hedges. *Pl. 47, fig.* 4.

Genus [345]. ELATER. *Lin*. Ten species.

E. sanguineus, *Lin*. Black; elytra of a bright red; about six lines; New Forest, not uncommon. *Pl. 47, fig.* 5.

Genus [346]. PROSTERNON. *Latr*. One species.

P. holosericeus, *Fab*. Brownish black, slightly metallic; densely covered above with a close silky pubescence, irregularly waved; about five or six lines; hedges and woods, not uncommon. *Pl. 47, fig.* 6.

Genus [347]. AGRYPNUS. *Esch*. Two species?

A. murinus, *Lin*. Brownish black; pubescent, and the pubescence mottled; from six to eight lines; very common, especially on sandy heaths. *Pl. 47, fig.* 7.

Genus [348]. HYPOLITHUS. *Esch*. Two species.

H. riparius, *Fab*. Brownish black, with a metallic tinge; basal joints of the antennæ and legs reddish; about three lines; meadows and pastures, rare. *Pl. 47, fig.* 8.

Genus [349]. CRYPTOHYPNUS. *Esch*. Four species.

C. Agricola, *Gyll*. Black; pubescent; each elytron with a pale oval spot at the humeral angles, and another not quite so large towards the apex; antennæ and legs testaceous; about two lines; meadows and pastures, not uncommon. *Pl. 47, fig.* 9.

Genus [350]. MELANOTUS. *Esch.* One species.

    M. fulvipes, *Herbst.* Deep pitchy black ; legs usually ful-
vous ; from seven to ten lines ; common in grassy places
and woods. *Pl. 47, fig.* 10.

Genus [351]. LUDIUS. *Latr.* One species.

    L. ferrugineus, *Lin.* Ferruginous brown, excepting the
head, antennæ, posterior edge and angles of the thorax,
scutellum and legs, which are brownish black ; eight to
eleven lines ; rare ; Richmond Park, Darenth Wood and
Windsor Forest. *Pl.* 48, *fig.* 1.

Genus [352]. CTENICERUS. *Latr.* Six species. The antennæ are merely serrated in the female.

    C. pectinicornis, *Lin.* Metallic green, somewhat pubescent;
antennæ and legs black ; about eight lines ; not uncommon in grassy places in the West and North of England.
Pl. 48, *fig.* 2.

Genus [353]. SELATOSOMUS. *Steph.* Two species?

    S. æneus, *Lin.* Either of a rich and shining brassy green
or of a metallic blue or violet ; antennæ pitchy ; legs
varying from fulvous to black ; about eight or nine lines ;
not uncommon in woods, very common in Scotland.
Pl. 48, *fig.* 3.

Genus [354]. CARDIOPHORUS. *Esch.* Three species.

    C. thoracicus, *Fab.* Black, with the thorax bright shining
red ; about four lines ; rare ; woods. *Pl.* 48, *fig.* 4.

Genus [355]. APLOTARSUS. *Steph.* Three species.

    A. rufipes, *Fab.* Obscure pitchy black ; legs either ful-
vous or pitchy ; about four lines ; meadows and pastures.
Pl. 48, *fig.* 5.

Genus [356]. SYNAPTUS. *Esch.* (CTENONYCHUS, *Steph.*)
One species.

    S. filiformis, *Fab.* (hirsutus, *Steph.*) Brownish black ;
thickly covered with long mouse-coloured hair ; antennæ
and legs varying from testaceous to pitchy ; about five or
six lines ; Bristol ? *Pl.* 48, *fig.* 6.

Genus [357]. ATHOÜS. *Esch.* Nine species ?

    A. vittatus, *Fab.* Pale pitchy black ; pubescent ; posterior
angles of the thorax, elytra (excepting their lateral margins), base of antennæ and legs all testaceous ; about
five lines ; New Forest and Hampstead, not common.
Pl. 48, *fig.* 7.

Genus [358]. CAMPYLUS. *Fischer.* One species.

    C. linearis, *Lin.* Either testaceous, with the head and
thorax red, or the elytra black and the lateral margins
reddish, the thorax varying from black to reddish testaceous, or sometimes merely its disk ; about six or seven
lines ; not uncommon in hedges. *Pl.* 48, *fig.* 8.

TRIBE II. MALACODERMATA. SHUCK.

Family [47]. CEBRIONIDÆ. Leach.

Genus [359]. DASCILLUS. *Latr.* (ATOPA, *Payk.*) One species.
D. cervinus, *Lin.* Either a dusky mouse-colour or fuscous testaceous; antennæ and legs testaceous; about five lines; not uncommon in hedges. *Pl.* 49, *fig.* 1.

Family [48]. ELODIIDÆ. Shuck.

Genus [360]. SCIRTES. *Illig.* Two species?
S. hemisphæricus, *Lin.* Black; slightly pubescent and shining; base of antennæ and the tibiæ and tarsi testaceous; about two lines; marshy places, by sweeping. *Pl.* 49, *fig.* 2.
Genus [361]. ELODES. *Latr.* (CYPHON, *Payk.*) Seventeen species.
E. lividus, *Fab.* Livid testaceous; disk of thorax and apex of antennæ fuscous or blackish; nearly three lines; common in humid situations. *Pl.* 49, *fig.* 3.
Genus [362]. EUBRIA. *Ziegler.* One species.
E. palustris, *Germ.* Pitchy black and shining; base of antennæ, apex of tibiæ, and tarsi testaceous; about one line; marshy places near Edinburgh and Oxford. *Pl.* 49, *fig.* 4.

Family [49]. LAMPYRIDÆ. Leach.

Genus [363]. DRILUS. *Oliv.* One species.
D. flavescens, *Oliv.* Male black, with the elytra yellowish testaceous and pubescent. The female subcylindrical, fleshy, of a bright orange, with an obscure marking on each side of each segment above. Male, about three lines—female, about ten lines; male common in woods and grassy places; both sexes parasitic in the larva state upon snails. *Pl.* 49, *fig.* 7 and 8.
Genus [364]. LAMPYRIS. *Lin.* One species.
L. noctiluca, *Lin.* Fuscous; margins of the thorax pale; from six to eight lines; grassy places; not uncommon in Kent. *Pl.* 49, *fig.* 5 and 6.
Genus [365]. DICTYOPTERUS. *Latr.* One species.
D. minutus, *Fab.* Intense black; pubescent; elytra of a rich blood red, and from their sculpture appearing

G

reticulated ; apex of antennæ fulvous ; about four lines ;
woods, in the South of England ; rare. *Pl.* 50, *fig.* 1.

Family [50]. TELEPHORIDÆ. Leach.

Genus [366]. SILIS. *Megerle.* One species. The male has
the antennæ serrated, and the lateral emargination of the
thorax deeper.
    S. ruficollis, *Fab.* Black and shining ; thorax red testa-
ceous ; antennæ and legs sometimes pitchy ; about four
lines ; grassy places, especially near marshes; not common. *Pl.* 50, *fig.* 2.
Genus [367]. TELEPHORUS. *De Geer.* Twenty-eight species?
    T. fuscus, *Lin.* Black ; thorax, excepting the anterior
margin, and the three basal joints of the antennæ red ;
legs obscure pitchy ; about eight lines ; hedges ; not
uncommon. *Pl.* 50, *fig.* 3.
Genus [368]. PODABRUS. *Fischer.* One species.
    P. alpinus, *Payk.* Testaceous ; thorax with an obscure
discoidal spot ; elytra either testaceous or slaty ; posterior legs and all the tarsi obscure ; about seven lines ;
trees in woods ; not uncommon. *Pl.* 50, *fig.* 4.
Genus [369]. RAGONYCHA. *Esch.* Five species.
    R. melanura, *Fab.* Reddish testaceous ; antennæ and
tarsi obscure ; apex of the elytra black ; about four
lines ; common. *Pl.* 50, *fig.* 5.
Genus [370]. MALTHINUS. *Latr.* Nineteen species.
    M. flaveolus, *Payk.* Yellow testaceous ; forehead blackish ;
disc of the thorax with two obscure spots ; elytra fuscous,
bright yellow at the apex, and the antennæ obscure at
the tip ; about three lines ; trees, by beating ; not uncommon. *Pl.* 50, *fig.* 6.

Family [51]. MELYRIDÆ.

Genus [371]. MALACHIUS. *Fab.* Fifteen species. In the
female the antennæ are simple.
    M. æneus, *Lin.* Metallic green ; head yellow in front ;
anterior angles of the thorax and the elytra vermilion
red, with an elongate triangular spot at the base, ex-
tending about half way down the suture, green ; about
four and a half lines ; not uncommon in meadows and
gardens. *Pl.* 51, *fig.* 1.
Genus [372]. APLOCNEMUS. *Steph.* One species.
    A. impressus, *Marsh.* Bronzy, with a pale pubescence ;

legs varying from fulvous to metallic black; about three lines; flowers, especially of the whitethorn. *Pl.* 51, *fig.* 2.

Genus [373]. DASYTES. *Payk.* Seven species.

D. cæruleus, *Fab.* Green or blue; antennæ black; about three lines; on the coasts, on flowers. *Pl.* 51, *fig.* 3.

Genus [374]. DOLICHOSOMA. *Steph.* One species.

D. linearis, *Fab.* Pale opaque green; pubescent; antennæ and legs metallic black; about three lines; the eastern coasts; not common. *Pl.* 51, *fig.* 4.

TRIBE III. SARCOTROGI. SHUCK.

Family [52]. TILLIDÆ. Leach.

Genus [375]. TILLUS. *Oliv.* One species.

T. elongatus, *Lin.* Bluish black; pubescent; thorax of a pale vermilion red; the male hitherto regarded as another species by the name of *T. ambulans, Fab.* is entirely black; about four lines; hedges and felled timber in woods. *Pl.* 52, *fig.* 1.

Genus [376]. TILLOIDEA. *Laporte.* One species.

T. unifasciata, *Oliv.* Black; pubescent; elytra red from the base about one-third of their length, and a yellowish white transverse fascia just beyond the middle; about three lines; Camberwell, on palings. *Pl.* 52, *fig.* 2.

Genus [377]. OPILUS. *Latr.* Two species.

O. mollis, *Lin.* Fuscous; pubescent; forehead, anterior margin of the thorax and the elytra, with an oblique fascia at the base, another beyond the middle, and the apex, and also the legs, all testaceous; antennæ pale red; about four or four and a half lines; hedges and beneath bark. *Pl.* 52, *fig.* 3.

Genus [378]. THANASIMUS. *Latr.* One species.

T. formicarius, *Lin.* Red; pubescent; head, anterior margin of the thorax and elytra black; the latter red at the base, and having two white bands, the anterior one angulated and placed just behind the basal red marking, the other towards the apex, less angulated and much broader; antennæ fuscous at the apex, and the tarsi pitchy; about four lines; beneath bark and on felled timber. *Pl.* 52, *fig.* 4.

Genus [379]. CLERUS. *Geoffr.* Two species.

C. apiarius, *Lin.* Of a rich blue; slightly pubescent; thorax slightly punctured and very shining; elytra red, with two bands and the apex blue; antennæ and tarsi

fuscous; about six lines; very rare; hedges, Dover, Norfolk, &c. *Pl.* 52, *fig.* 5.
Genus [380]. NECROBIA. *Oliv.* Four species.
  N. ruficollis, *Oliv.* Blue, shining; pubescent; thorax, base of elytra and legs red; antennæ black; about two and a half lines; bones and carrion; not uncommon. *Pl.* 52, *fig.* 6.
Genus [381]. CORYNETES. *Payk.* One species.
  C. violaceus, *Lin.* Bright shining blue; pubescent; antennæ black, the base paler; about two lines; carrion; common. *Pl.* 52, *fig.* 7.

TRIBE IV. HYLEPHAGI. SHUCKHARD.

Family [53]. BOSTRICHIDÆ. Leach.

Genus [382]. SPHINDUS. *Dejean.* One species.
  S. dubia, *Gyll.* (Gyllenhallii, *Dej.*) Brown black, shining; antennæ and legs red; one line and a half; beneath bark; Sherwood Forest; rare. *Pl.* 53, *fig.* 1.
Genus [383]. CIS. *Latr.* Thirteen species.
  C. Boleti, *Fab.* Chesnut brown; slightly pubescent; antennæ and legs fulvous; nearly two lines; in Boleti; common. *Pl.* 53, *fig.* 2.
Genus [384]. BOSTRICHUS. *Geoffr.* One species.
  B. capucinus, *Lin.* Opaque black; slightly pubescent; elytra and tarsi red; about five lines; woods and forests; not common. *Pl.* 53, *fig.* 3.
Genus [385]. APATE. *Fab.* One species.
  A. sinuata, *Fab.* Pitchy black or chesnut, shining; tarsi reddish; about three lines; Epping Forest and Hackney; rare. *Pl.* 53, *fig.* 4.
Genus [386]. DINODERUS. *Steph.* One species.
  D. substriatus, *Payk.* Pitchy black, opaque; elytra brown; club of antennæ and legs reddish. *Pl.* 53, *fig.* 5.

Family [54]. ANOBIIDÆ. Shuck.

Genus [387]. ANOBIUM. *Fab.* Ten species.
  A. striatum, *Oliv.* Pitchy brown, opaque; antennæ and legs reddish; nearly two lines; old timber; common. *Pl.* 53, *fig.* 6.
Genus [388]. DRYOPHILUS. *Chevrolat.* Three species.
  D. pusillus, *Gyll.* (Anobioides, *Guer.*) Pitchy brown; antennæ and legs testaceous; about one line and a half; stumps of broom; Coombe Wood. *Pl.* 53, *fig.* 7.

Genus [389]. DORCATOMA. *Herbst.* Three species.
> D. Dresdense, *Herbst.* Black; legs and antennæ reddish; eyes entire; about one line; decayed trees; Surrey and Suffolk. *Pl.* 53, *fig.* 8.

Family [55]. PTILINIDÆ. Shuck.

Genus [390]. LASIODERMA. *Steph.* One species.
> L. testaceum, *Steph.* Opaque testaceous; pubescent; legs and antennæ pale; about a line; old houses. *Pl.* 54, *fig.* 1.

Genus [391]. OCHINA. *Ziegler.* One species.
> O. Ptinoides, *Marsh.* Chesnut brown; pubescent; elytra with two ashy pale bands, legs and antennæ reddish; about one line and a half; old dry stumps of ivy; not common. *Pl.* 54, *fig.* 2.

Genus [392]. XILETINUS. *Latr.* Two species?
> X. pectinatus, *Fab.* Opaque black, clothed with silken pile; legs and antennæ red; about a line and a half; old trees. *Pl.* 54, *fig.* 3.

Genus [393]. PTILINUS. *Fab.* Two species.
> P. pectinicornis, *Lin.* Opaque fuscous black; elytra brown; antennæ and legs testaceous; about two lines; old wood. *Pl.* 54, *fig.* 4.

Family [56]. PTINIDÆ. Leach.

Genus [394]. HEDOBIA. *Latr.* One species.
> H. imperialis, *Lin.* Fuscous; pubescent; the sides of the thorax, the suture of the elytra (dilated at the base), two lateral spots at about two-thirds of their length, and their apex narrowly, all whitish; antennæ and legs fuscescent; about two and a half lines; hedges; not uncommon. *Pl.* 54, *fig.* 5.

Genus [395]. PTINUS. *Lin.* Seven species.
> P. Fur, *Lin.* Dirty reddish-brown; slightly pubescent; legs and antennæ fulvescent; the female is more robust and has a couple of interrupted whitish bands across the elytra; about two lines; houses, &c.; not uncommon. *Pl.* 54, *fig.* 6.

Genus [396]. MEZIUM. *Leach.* One species.
> M. sulcatum, *Fab.* Head, antennæ, thorax and legs covered with ashy scales; elytra bright shining chesnut; about a line and a half; lumber rooms in old houses; not common. *Pl.* 54, *fig.* 7.

Genus [397]. GIBBIUM. *Scopoli.* (1777). One species.
  G. Scotias, *Fab.* Bright shining chesnut; glabrous; an-
    tennæ and legs clothed with dirty yellowish scales;
    about one line; lumber cupboards in old houses; not
    common. *Pl.* 54, *fig.* 8.

TRIBE V. XYLOTROGI. LATR.

Family [57]. LYMEXYLONIDÆ. Steph.

Genus [398]. HYLECÆTUS. *Latr.* In the female the palpi
  are simple. One species.
  H. dermestoides, *Lin.* Very variable in colour, but usually
    black, with the elytra and legs testaceous, the apex of
    the former black; three and a half to seven lines; Sher-
    wood Forest, Notts; rare. *Pl.* 51, *fig.* 6.
Genus [399]. LYMEXYLON. *Fab.* In the female the palpi
  are simple. One species.
  L. navale, *Lin.* Variable in colour, but usually testaceous,
    with the head, antennæ and apex of the elytra black; or
    with the thorax and whole elytra more or less clouded
    with black; three and a half to six lines; Windsor Fo-
    rest; very rare. *Pl.* 51, *fig.* 5.

SECTION II. HETEROMERA. LATREILLE.

TRIBE I. TRACHELIDES. LATR.

Family [58]. PYROCHROIDÆ. Leach.

Genus [400]. PYROCHROA. *Geoffr.* Two species.
  P. coccinea, *Lin.* Atrous; thorax above and the elytra of
    a bright blood-red; about seven lines; woods in Kent;
    not common. *Pl.* 55, *fig.* 1.

Family [59]. LAGRIIDÆ. West.

Genus [401]. LAGRIA. *Fab.* One species.
  L. hirta, *Lin.* Black, shining; very pubescent; elytra of
    a dirty pale testaceous; about four lines; common, on
    flowers. *Pl.* 55, *fig.* 2.

47

Family [60]. XYLOPHILIDÆ. Shuck.

Genus [402]. XYLOPHILUS. *Latr.* In the female the antennæ
are shorter, and the second and third joints are nearly equal.
One species.
    X. oculatus, *Payk.* Deep black; with the antennæ, elytra
    and legs dark testaceous; about one line and a half;
    from willows, by beating; not uncommon. *Pl.* 55,
    fig. 3.
Genus [403]. ADERUS. *West.* In the male the antennæ are
longer. One species.
    A. populneus, *Fab.* Red testaceous, covered with a close
    sericeous pubescence; with a waved band across the
    middle which extends laterally upwards towards the
    shoulders, and the apex broadly denuded; eyes black;
    about a line and a half; by sweeping in the air on sultry
    summer evenings in the vicinity of ivy; rare. *Pl.* 55,
    fig. 4.

Family [61]. NOTOXIDÆ. Steph.

Genus [404]. NOTOXUS. *Illig.* Two species.
    N. Monoceros, *Lin.* Variable in colour, but usually red-
    dish-testaceous; elytra with black spots disposed in
    pairs, or sometimes nearly entirely black; about three
    lines; in sandy pastures, by sweeping. *Pl.* 55, *fig.* 5.
Genus [405]. ANTHICUS. *Fab.* Twelve species.
    A. antherinus, *Lin.* Pitchy black, slightly shining; elytra
    with two humeral and one central common spot red;
    tibiæ and tarsi reddish testaceous; about two lines;
    hedges, by beating. *Pl.* 55, *fig.* 6.

Family [62]. MORDELLIDÆ. Leach.

Genus [406]. MORDELLA. *Lin.* Ten species.
    M. fasciata, *Fab.* Deep black; elytra variegated with
    bright ashy down; from four to five lines; on flowers,
    in woods; not common. *Pl.* 56, *fig.* 1.
Genus [407]. ANASPIS. *Geoffr.* Eighteen species.
    A. frontalis, *Lin.* Black; mouth, base of antennæ, fore-
    head and anterior legs pale testaceous; about two lines;
    flowers in the spring; very common. *Pl.* 56, *fig.* 2.
Genus [408]. RHIPIPHORUS. *Fab.* One species.
    R. paradoxus. *Lin.* Variable in colour; usually either

testaceous, with the head, antennæ, centre of the thorax, pectus, legs and tips of the elytra black—or black, with the posterior angles of the thorax and abdomen testaceous; about seven lines; flowers and wasps' nests, not uncommon in the latter. *Pl.* 56, *fig.* 3.

Family [63]. CANTHARIDÆ. Leach.

Genus [409]. SITARIS. *Latr.* One species.
    S. humeralis, *Fab.* Black; elytra yellowish testaceous at the base; about six lines; parasitic on *Osmiæ*; Chelsea; very rare. *Pl.* 56, *fig.* 4.

Genus [410]. SYBARIS. *Steph.* One species.
    S. immunis, *Marsh.* Fuscous; head, thorax, elytra and legs testaceous; the knees and tarsi of the latter black; five and a half lines; Rochford, Essex; very rare. *Pl.* 56, *fig.* 5.

Genus [411]. CANTHARIS. *Geoffr.* One species.
    C. vesicatoria, *Lin.* Bright shining golden green or coppery; antennæ black; tarsi bluish; seven to ten lines; on the ash; Essex and near Southampton, in great profusion at times. *Pl.* 56, *fig.* 6.

Genus [412]. MELOË. *Lin.* Nine species.
    M. violaceus, *Marsh.* Of a bright violet blue; from six lines to an inch long; meadows and heaths in the early spring and autumn; common. *Pl.* 56, *fig.* 7.

TRIBE II. MELASOMA. LATR.

Family [64]. BLAPSIDÆ. Leach.

Genus [413]. BLAPS. *Fab.* Three species.
    B. fatidica, *Sturm.* Entirely of a deep bright black; nine to twelve lines; cellars and outhouses; not common. *Pl.* 57, *fig.* 1.

Genus [414]. PEDINUS. *Latr.* One species.
    P. femoralis, *Lin.* Black, slightly shining; tip of antennæ and tarsi pitchy; four lines and a half; sandy places, Deal; very rare. *Pl.* 57, *fig.* 2.

Genus [415]. HELIOPHILUS. *Dejean?* Two species?
    H. obsoletus, *Marsh.* Entirely black, slightly shining; five lines; coast of Suffolk; very rare. *Pl.* 57, *fig.* 3.

Family [65]. OPATRID.E. Shuck.

Genus [416]. PHYLAN. *Megerle.* One species.
  P. gibbus, *Fab.* Entirely of a shining black; about four
  lines; sandy coasts; not uncommon. *Pl.* 57, *fig.* 4.
Genus [417]. OPATRUM. *Fab.* Two species.
  O. sabulosum, *Lin.* Of a dull dirty black; opaque; antennæ pitchy; about four lines; sandy situations, especially the coasts. *Pl.* 57, *fig.* 5.
Genus [418]. CRYPTICUS. *Latr.* One species.
  C. quisquilius, *Lin.* Shining black; antennæ, legs and
  tarsi pitchy; about three lines; sandy situations on the
  coasts; common. *Pl.* 57, *fig.* 6.

Family [66]. TENEBRIONIDÆ. Steph.

Genus [419]. ALPHITOBIUS. *Steph.* One species.
  A. picipes, *Steph.* Pitchy black; antennæ and legs red-
  dish; two lines and a half; under turf and in flour; not
  common. *Pl.* 58, *fig.* 1.
Genus [420]. ULOMA. *Megerle.* The female has the mandi-
  bles simple. One species.
  U. cornuta, *Fab.* Of a dull opaque red; eyes black;
  about two lines; in flour; not common. *Pl.* 58, *fig.* 2.
Genus [421]. TENEBRIO. *Lin.* Four species?
  T. molitor, *Lin.* Entirely of an obscure pitchy black,
  slightly shining; about six or seven lines; in flour.
  Pl. 58, *fig.* 2.
Genus [422]. STENE. *Kirby.* One species.
  S. ferruginea, *Fab.* Of a dull opaque red; nearly two
  lines; in flour. *Pl.* 58, *fig.* 3.

TRIBE III. TAXICORNES. LATR.

Family [67]. HYPOPHLÆIDÆ. Shuck.

Genus [123]. HYPOPHLÆUS. *Fab.* Four species.
  H. castaneus, *Fab.* Bright shining chesnut; antennæ and
  legs a little paler; about three lines; beneath bark;
  Sherwood Forest. *Pl.* 58, *fig.* 5.
Genus [424]. BOLITOPHAGUS. *Fab.* One species.
  B. Agricola, *Fab.* Of an opaque pitchy brown; antennæ
  and legs reddish; nearly two lines; Boleti in woods.
  Pl. 58, *fig.* 6.

H

Genus [425]. ALPHITOPHAGUS. *Steph.* One species.
A. quadripustulatus, *Davis.* Shining reddish or pale testaceous; elytra with the suture and two bands black;
antennæ and legs pale; about one line and a half; in
flour. *Pl. 58, fig. 7.*

Family [68]. SARROTRIIDÆ. Shuck.

Genus [426]. SARROTRIUM. *Illiger.* One species.
S. muticum, *Lin.* Entirely of an opaque obscure black;
about two lines; sandy heaths; not common. *Pl. 58,
fig. 8.*

Family [69]. DIAPERIDÆ. West.

Genus [427]. PHALERIA. *Latr.* One species.
P. cadaverina, *Fab.* Testaceous, slightly shining; disk of
the elytra with a blackish or pitchy spot; antennæ and
legs pale; about three lines; sandy coasts. *Pl. 59,
fig. 1.*
Genus [428]. DIAPERIS. *Geoffr.* One species.
D. Boleti, *Lin.* Bright shining black; elytra with two
waved yellow transverse bands and a spot at the apex of
the same colour; about four lines; in Boleti; very rare.
Pl. 59, fig. 2.
Genus [429]. PLATYDEMA. *Laporte.* Three species?
P. ænea, *Payk.* Bronzy black; base of antennæ and legs
pitchy red; two lines and a half; stumps of trees; not
uncommon, but local. *Pl. 59, fig. 3.*

TRIBE IV. STENELYTRA. LATR.

Family [70]. HELOPIDÆ. Steph.

Genus [430]. HELOPS. *Fab.* Four species.
H. cæruleus, *Fab.* Of a rich blue black; apex of the antennæ and the tarsi reddish brown; seven to nine lines;
hollow trees, especially in gardens; not uncommon.
Pl. 59, fig. 4.

Family [71]. CISTELIDÆ. Leach.

Genus [431]. ERYX. *Steph.* One species.
E. niger, *De Geer.* Entirely of a shining black; antennæ

and legs pitchy; six to eight lines; under bark and in hollow willows; Windsor; not common. *Pl.* 59, *fig.* 5.

Genus [432]. MYCETOCHARUS. *Latr.* One species.

    M. scapularis, *Gyll.* Pitchy black and shining; basal joints of the antennæ, a humeral spot on the elytra, and the tibiæ and tarsi fulvous; about three lines; hollow willows; not common. *Pl.* 59, *fig.* 6.

Genus [433]. CISTELA. *Fab.* Five species.

    C. ceramboides, *Lin.* Black; with the elytra testaceous; nearly six lines; flowers in June; not common. *Pl.* 59, *fig.* 7.

Genus [434]. OMOPHLUS. *De Jean.* One species.

    O. Amerinæ, *Curtis.* Black, shining; elytra testaceous; apex of the tibiæ and the tarsi pitchy red or chesnut; about four lines; rare; Isle of Portland. *Pl.* 59, *fig.* 8.

Genus [435]. CTENIOPUS. *Solier.* One species.

    C. sulphurea, *Lin.* Of a bright rich sulphur yellow; antennæ and tibiæ pale reddish; the apex of the former and the tarsi pitchy; about four lines; flowers on the coasts; common, but local. *Pl.* 60, *fig.* 1.

Family [72]. NOTHIDÆ. Shuck.

Genus [436]. NOTHUS. *Megerle.* Two species?

    N. bipunctatus, *Fab.* Reddish testaceous; with the head, the apex of the antennæ, two spots on the thorax, knees of the femora, and the four posterior tarsi black; four to six lines; not uncommon in Monk's Wood, Hunts. *Pl.* 60, *fig.* 2.

Genus [437]. CONOPALPUS. *Gyll.* Two species.

    C. testaceus, *Oliv.* Pale testaceous; apex of the antennæ brown; about three lines and a half; flowers in June; not common. *Pl.* 60, *fig.* 3.

Family [42]. MELANDRYIDÆ. Leach.

Genus [438]. MELANDRYA. *Fab.* Two species.

    M. caraboides, *Lin.* Black and shining, with a slight metallic greenish or bluish reflection; apex of the antennæ and of the tarsi red; five to eight lines; trees and palings; common. *Pl.* 60, *fig.* 4.

Genus [439]. PHLOIOTRYA. *Steph.* One species.

    P. rufipes, *Gyll.* Pitchy brown; base of antennæ, palpi and legs testaceous red; four to eight lines; rotten trees;

not common ; Windsor, Sherwood, and New Forests.
Pl. 60, *fig.* 5.
Genus [440]. DIRCÆA. *Fab.* One species.
    D. variegata, *Fab.* Fuscous; antennæ and elytra testa-
ceous, the latter mottled with fuscous; about three lines;
old trees; Windsor; very rare. *Pl.* 60, *fig.* 6.
Genus [441]. HYPULUS. *Payk.* One species.
    H. quercinus, *Payk.* Pitchy red ; thorax dark ; antennæ
and elytra dull red ; the latter with a curved line at the
base, another waved about the middle, and the apex
black ; legs fulvous; nearly four lines ; old trees ; rare.
Pl. 60, *fig.* 7.
Genus [442]. ABDERA. *Steph.* Three species?
    A. quadrifasciata, *Curt.* Pitchy brown ; anterior and
posterior margins of the thorax and two waved bands on
the elytra, as well as the antennæ and legs, testaceous;
about two lines ; flowers in hedges ; rare. *Pl.* 60, *fig.* 8.
Genus [443]. SCRAPTIA. *Latr.* Two species.
    S. nigricans, *Steph.* Pitchy fuscous, covered with a golden
pubescence ; antennæ fuscous; legs testaceous ; one line
and a half; flowers in gardens; Surrey; rare. *Pl.* 61,
fig. 1.
Genus [444]. HALLOMENUS. *Payk.* Three species.
    H. dimidiatus, *Steph.* Brown, covered with a silky pile ;
elytra pale brown ; legs testaceous ; a little more than
two lines ; North of England ; rare. *Pl.* 61, *fig.* 2.
Genus [445]. ORCHESIA. *Latr.* Three species.
    O. micans, *Illig.* Pitchy brown, with a close silken pubes-
cence ; elytra a little paler ; antennæ, palpi and legs
testaceous ; about two lines ; in Boleti ; not uncommon.
Pl. 61, *fig.* 3.

Family [74]. ŒDEMERIDÆ. Leach.

Genus [446]. ISCHNOMERA. *Steph.* In the female of the species figured the antennæ have eleven joints. Four species.
    I. melanura, *Lin.* Head and thorax reddish testaceous;
elytra covered with a silken pile, pale testaceous, with
their apex black ; antennæ and legs testaceous, the middle of the former and the four posterior thighs fuscous ;
five to eight lines ; coasts, near timber. *Pl.* 61, *fig.* 4.
Genus [447]. ŒDEMERA. *Oliv.* In the female the legs are
simple. Three species.
    Œ. cærulea, *Lin.* Of a rich golden green, shining; an-
tennæ and tarsi black ; about five or six lines ; flowers ;
common. *Pl.* 61, *fig.* 5.

53

Genus [448]. Oncomera. *Steph.* The female has the legs simple. One species.

    O. femorata, *Fab.* Pale fuscous; sides of the thorax and a ring at the knees of the femora black; seven to nine lines; flying at night; Kent, Surrey and Berks; rare. *Pl.* 61, *fig.* 6.

Family [75]. Salpingidæ. Leach.

Genus [449]. Mycterus. *Clairv.* One species.

    M. griseus, *Clairv.* Black, with a greenish grey pile; about four lines; hedges; Devonshire; very rare. *Pl.* 61, *fig.* 7.

Genus [450]. Sphæriestes. *Kirby.* Five species.

    S. foveolatus, *Ljung.* Brassy black and shining; base of antennæ and tarsi reddish; about two lines and a half; under elms; Scotland; not common. *Pl.* 61, *fig.* 8.

      Note.—It is here that the genus Lissodema, Curt. should be placed, instead of among the *Engidæ*, for its type is an insect congeneric with the *Sphæriestes quadri pustulatus*, Marsh. which was the original type of Mr. Kirby's genus *Sphæriestes*, but the *Sph. niger* having been treated as such, and as there are evidently two forms of structure in the antennæ of the genus, it will be convenient to treat those in which the club is gradually formed, of which the type is the *niger*, and to which the one-figured belongs, as the genus *Sphæriestes*, and those in which the club distinctly consists of three joints, as *Lissodema*.

Genus [451]. Salpingus. *Illig.* Three species.

    S. ruficollis, *Lin.* Bright fulvous, shining; elytra bluish or green; head between the eyes and apex of the antennæ fuscous; about two lines; beneath bark and in hedges by beating; not common. *Pl.* 61, *fig.* 9.

Section III. TETRAMERA. Latreille.

Subdivision I. RHYNCHOPHORA. Latr.

Tribe I. SPURII. Schön.

Family [76]. Rhinomaceridæ. Shuck.

Genus [452]. Rhinomacer. *Fab.* One species.

    R. Attelaboides, *Fab.* Greenish, covered with grey pubescence; antennæ and legs testaceous; three lines; the North of Scotland; very rare. *Pl. 62, fig. 7.*

Family [77]. ANTHRIBIDÆ. Shuck.

Genus [453]. CHORAGUS. *Kirby.* Two species?
C. Sheppardi, *Kirby.* Brownish black; antennæ and legs pitchy; nearly a line; willows and grassy places, on heaths; very rare. *Pl. 62, fig. 2.*
Genus [454]. ANTHRIBUS. *Fab.* One species. In the female the antennæ are shorter.
A. albinus, *Lin.* Brown, covered with scales; the head above, two spots in the centre of the elytra, and their apex, white; the latter variegated with a few black tufts; the eighth and ninth joints of the antennæ also white; and the legs variegated; from four to six lines; within old wood; not common. *Pl. 62, fig. 3.*
Genus [455]. TROPIDERES. *Schön.* Two species.
T. niveirostris, *Oliv.* Blackish, with the head above, the scutellum, shoulders, and apex of the elytra and pygidium, white; antennæ reddish testaceous; the club darker; legs variegated with white; three lines; hedges not common. *Pl. 62, fig. 4.*
Genus [456]. PLATYRHINUS. *Clairv.* One species.
P. latirostris, *Bons.* Bluish black; the head above and the apex of the elytra dirty white; their surface and the legs also mottled with white and black; about six lines; old wood; Worcestershire. *Pl. 62, fig. 5.*
Genus [457]. BRACHYTARSUS. *Schön.* Two species.
B. scabrosus, *Fab.* Black; elytra red, variegated with black and whitish tufts; about two lines; beneath the bark of elms; not uncommon. *Pl. 62, fig. 6.*

Family [78]. BRUCHIDÆ. Leach.

Genus [458]. BRUCHUS. *Geoffr.* Twelve species?
B. Pisi, *Lin.* Black, mottled with white; the pygidium white, with two black spots; the base of the antennæ and the anterior legs testaceous; about two lines; common in the vicinity of pea fields. *Pl. 62, fig. 1.*

TRIBE II. GENUINI. Schön.

RACE I. ORTHOCERI. Schön.

Family [79]. ATTELABIDÆ. West.

Genus [459]. APODERUS. *Oliv.* One species.
A. Avellanæ, *Oliv.* Red; with the head, antennæ, scutel-

lum, knees and tarsi black, and sometimes the centre of the thorax ; about three lines ; common on the hazel. *Pl.* 63, *fig.* 1.

Genus [460]. ATTELABUS. *Lin.* One species.
    A. Curculionoides, *Lin.* Red ; with the head, antennæ, scutellum and legs black ; about three lines ; common on the hazel. *Pl.* 63, *fig.* 2.

Genus [461]. RHYNCHITES. *Herbst.* Eighteen species?
    R. Bacchus? *Lin.* Of a rich golden copper ; the rostrum bluish ; antennæ and tarsi black ; about four lines ; on the blackthorn ; very rare. *Pl.* 63, *fig.* 4.

Genus [462]. DEPORAUS. *Leach.* One species.
    D. Betulæ, *Lin.* Entirely of a deep shining black ; about two lines ; common in woods on the birch. *Pl.* 63, *fig.* 3.

Family [80]. APIONIDÆ. Shuck.

Genus [463]. APION. *Herbst.* Sixty-eight species?
    A. frumentarium, *Lin.* Of a uniform blood red ; about two lines ; common on the leaves of dock. *Pl.* 63, *fig.* 7.

Genus [464]. OXYSTOMA. *Leach.* Three species?
    O. Ulicis, *Forst.* Black, densely clothed with a silvery grey decumbent pubescence ; anterior legs and the base of the antennæ fulvous ; about a line and a half ; on the furze ; common. *Pl.* 63, *fig.* 6.

Family [81]. RHAMPHIDÆ. Shuck.

Genus [465]. RHAMPHUS. *Clairv.* One species.
    R. pulicarius, *Herbst.* Deep black and shining ; the base of the antennæ fulvous ; not quite a line ; common on the birch and willow. *Pl.* 63, *fig.* 5.

RACE II. GONATOCERI. Schön.

BAND I. MECORHYNCHI. Schön.

Family [82]. COSSONIDÆ. Shuck.

Genus [466]. RHYNCOLUS. *Germ.* Four species.
    R. truncorum, *Schüp.* Blackish brown, with a brassy tinge ; antennæ and legs piceous ; about a line and a half ; beneath bark ; Ireland. *Pl.* 64, *fig.* 1.

Genus [467]. COSSONUS. *Clairv.* Two species.
    C. linearis, *Fab.* Black, pitchy, or testaceous ; the antennæ and legs paler ; about three lines ; on willows ; not common. *Pl.* 63, *fig.* 8.

Family [83]. CALANDRIDÆ. Shuck.

Genus [468]. CALANDRA. *Clairv.* Two species.
C. granaria, *Lin.* Of a uniform pitchy black; antennæ and legs pitchy red; nearly two lines; in granaries. *Pl.* 64, *fig.* 2.

Family [84]. CIONIDÆ. Shuck.

Genus [469]. NANOPHYES. *Schön.* (Spherula, *Steph.*) One species.
N. Lythri, *Fab.* Black, with a greyish down; the elytra with a testaceous angulated band, the base of the antennæ, and the tibiæ and tarsi fulvous; about one line; on Lythrum salicariæ; not common. *Pl.* 64, *fig.* 3.
Genus [470]. MECINUS. *Germar.* Three species.
M. Pyraster, *Herbst.* (semicylindricus, *Marsh.*) Black, loosely covered with grey pile; about two lines; common in marshy meadows. *Pl.* 64, *fig.* 4.
Genus [471]. GYMNETRON. *Schön.* (and RHINUSA and MIARUS, *Steph.*) Eight species.
G. Beccabungæ, *Lin.* Black, covered with a dense ashy pubescence; base of the antennæ and the tibiæ and tarsi fulvous; about a line and a half; grassy places; not common. *Pl.* 64, *fig.* 5.
Genus [472]. CIONUS. *Clairv.* (and CLEOPUS, *Steph.*) Six species.
C. Scrophulariæ, *Lin.* Blue black; the thorax cream-coloured; the elytra with two common black marks on the suture, the one near the scutellum, and the other towards the apex, the former edged behind, and the latter in front, with white; and each elytron with four alternate rows of black and white tesselated spots; about two lines and a half; common on the Scrophularia in ditches. *Pl.* 65, *fig.* 1.

Family [85]. CRYPTORHYNCHIDÆ. Shuck.

Genus [473]. OROBITIS. *Germ.* One species.
O. cyaneus, *Lin.* Bright blue black, shining; antennæ and legs pitchy; about a line and a half; grassy places; not common. *Pl.* 64, *fig.* 2.
Genus [474]. RHYTIDOSOMA. *Steph.* One species.
R. globula, *Herbst.* Black and shining; the scutellum white; about one line; willows, marshy places. *Pl.* 65, *fig.* 3.

Genus [475]. POOPHAGUS. *Schön.* Two species.
> P. Sisymbrii, *Fab.* Brownish black, densely clothed with ashy scales, and having several denuded spots on the thorax and elytra; about two lines; not uncommon on the Sisymbrium amphibium in marshy places. *Pl.* 65, *fig.* 4.

Genus [476]. RHINONCHUS. *Schön.* Eight species?
> R. pericarpius, *Lin.* Brown black; scutellum pale; antennæ and legs pitchy red; about two lines; common on thistles. *Pl.* 65, *fig.* 5.

Genus [——]. NEDYUS. (*Schön.* formerly.) *Steph.* Thirty-three species.
> N. pollinarius. *Forst.* Brownish black, maculated with ashy scales; antennæ and legs dark pitchy; about two lines; common on nettles. *Pl.* 65, *fig.* 6.

Genus [477]. CEUTORHYNCUS. *Schüp.* Eleven species.
> C. Quercus, *Herbst.* Blackish or reddish brown; elytra undulated with ashy scales; about one line and a quarter; common on oaks. *Pl.* 65, *fig.* 7.

Genus [478]. ACALLES. *Schön.* Four species.
> A. Ptinoides, *Marsh.* Brown black; the setæ and a transverse line of tufts in the centre of the elytra deep black, the apex of the latter lutescent; one line and a half; sandy heaths; not common. *Pl.* 65, *fig.* 8.

Genus [479]. MONONYCHUS. *Schüp.* One species.
> M. Pseudacori, *Fab.* Entirely of an opaque black; about two lines; in the seed pods of the Iris; Isle of Wight; local, but not uncommon. *Pl.* 66, *fig.* 1.

Genus [480]. CŒLIODES. *Schön.* One species.
> C. Geranii, *Payk.* Black, with grey scales; about one line and a half; widely distributed, but not common; on the Geranium pratense. *Pl.* 66, *fig.* 2.

Genus [481]. CRYPTORHYNCHUS. *Illig.* One species.
> C. Lapathi, *Lin.* Black, variously mottled with white and brown scales and black tufts; an angulated whitish band proceeding from the shoulders and the posterior portion of the elytra also whitish; femora with alternate white and black rings; about four lines and a half; common on willows in marshy places. *Pl.* 66, *fig.* 3.

Family [86]. BARIDIDÆ. Shuck.

Genus [482]. BARIDIUS. *Schön.* (BARIS. *Steph.*) Five species.
> B. T. album, *Lin.* (Atriplicis, *Steph.*) Bright shining black, covered beneath with white scales; the plane interstices on the elytra properly covered with lines of

I

light scales ; about two lines ; humid meadows ; Battersea, by sweeping. *Pl. 66, fig. 4.*

Family [87]. ERIRHINIDÆ. Shuck.

Genus [483]. BAGOUS. *Germ.* Four species.
    B. binotatus, *Steph.* Black, with brownish scales, and each elytron, just beyond the middle, having a pale spot; about a line and a half; humid meadows; Battersea. *Pl. 66, fig. 5.*
Genus [484]. LYPRUS. *Schön.* One species.
    L. cylindrus, *Gyll.* Densely covered with ashy scales ; about one line and a quarter ; very rare ; on the water-cresse. *Pl. 66, fig. 6.*
Genus [485]. ORTHOCHÆTES. *Müller.* One species.
    O. setiger, *Germ.* Reddish ; the elevated interstices of the elytra having a row of erect setæ; about one line ; rare. *Pl. 66, fig. 7.*
Genus [486]. TACHYERGES. *Schön.* Five species.
    T. stigma, *Germ.* Entirely bright black ; the scutellum snowy white; about a line and a half; marshy places ; not uncommon. *Pl. 66, fig. 8.*
Genus [487]. ORCHESTES. *Illig.* Fifteen species?
    O. Quercus, *Lin.* Reddish testaceous ; the elytra with a large triangular cinereous spot in front ; about two lines ; common on the oak. *Pl. 67, fig. 1.*
Genus [488]. ANOPLUS. *Schön.* One species.
    A. plantaris, *Nœzen.* Bright shining black, scutellum whitish ; nearly one line ; by beating the birch; not common. *Pl. 67, fig. 2.*
Genus [489]. PACHYRHINUS. *Kirby.* (PHYTOBIUS. *Schön.*) Eight species.
    P. Myriophylli, *Gyll.* Black, densely clothed with lutescent scales ; legs testaceous ; about a line and a half; on the water dock; not common. *Pl. 67, fig. 3.*
Genus [490]. SIBYNES. *Schön.* Three species.
    S. arenariæ, *Kirby.* Densely clothed with brownish satiny scales; the elytra with a darker spot gradually expanding from the scutellum and abruptly truncated; this edged with silvery grey; antennæ and legs testaceous; about a line ; sandy places ; Hampstead. *Pl. 67, fig. 4.*
Genus [491]. MICCOTROGUS. *Schön.* Two species?
    M. picirostris, *Schön.* Pitchy red, densely covered with ashy scales ; the base of the antennæ, apex of the rostrum, and the tibiæ and tarsi red or testaceous ; about a line ; grass, beneath fir trees; not uncommon. *Pl. 67, fig. 5.*

59

Genus [492]. Tychius. *Germ.* Eight species.
   T. venustus, *Fab.* Black, covered with brownish scales,
   with three longitudinal pale lines, one central and two
   lateral; tibiæ and tarsi rufo-ferruginous; about two
   lines; sandy heaths. *Pl. 67, fig. 6.*
Genus [493]. Amalus. *Schön.* Two species?
   A. Scortillum, *Herbst.* Black; the base of the suture with
   ashy scales; apex of elytra and legs red; about one line;
   marshy places. *Pl. 67, fig. 7.*
Genus [494]. Balaninus. *Germ.* Ten species.
   B. Nucum, *Lin.* Densely covered with yellowish or ashy
   scales, which are mottled on the elytra; rostrum, antennæ and legs pitchy red; about four or five lines, in-
   cluding the rostrum; on the nut; common. *Pl. 67, fig. 8.*
Genus [495]. Anthonomus. *Germ.* Eight species.
   A. Ulmi, *De Geer.* Reddish testaceous; thorax with a
   white central longitudinal line; the scutellum and a posterior band on the elytra white; the shoulders mottled;
   about two lines; flowers, in hedges. *Pl. 68, fig. 1.*
Genus [496]. Ellescus. *Megerle.* One species.
   E. bipunctatus, *Lin.* Black, densely covered with ashy
   scales; the elytra, just beyond the centre, having two
   denuded spots; tibiæ and tarsi reddish; about a line
   and a half; Scotland; not common. *Pl. 68, fig. 2.*
Genus [497]. Hydronomus. *Schön.* One species.
   H. Alismatis, *Marsh.* Black, variegated with whitish ashy
   scales; tibiæ testaceous; about two lines; brooks, on
   the Alismatis plantago; not uncommon. *Pl. 68, fig. 3.*
Genus [498]. Grypidius. *Schön.* One species.
   G. equiseti, *Fab.* Black, variegated with whitish scales;
   the elytra with their margin and two discoïdal spots
   white; about four lines; on the horse-tail grass, in
   marshy places; not uncommon. *Pl. 68, fig. 4.*
Genus [499]. Erirhinus. *Schön.* Three species.
   E. Nereis, *Payk.* Fusco-piceous, densely covered with
   ashy scales; elytra with a common pitchy spot in front;
   legs ferruginous; about two lines; marshy places.
   Pl. 68, fig. 5.
Genus [——]. Dorytomus. *Steph.* (Erirhinus. *Schön.*)
   Nine species.
   D. vorax, *Herbst.* (longimanus, *Forst.*) Pitchy black,
   densely covered and variegated with ashy scales; an-
   tennæ and legs ferruginous; about three lines; willows,
   marshy places. *Pl. 68, fig. 6.*
Genus [—]. Notaris. *Steph.* (Erirhinus. *Schön.*) Four
   species.
   N. bimaculatus, *Fab.* Black, opaque, densely covered with

fuscous scales; the elytra with two white pilose spots just beyond the middle; about four lines and a half; marshy places. *Pl.* 68, *fig.* 7.

Genus [500]. THAMNOPHILUS. *Schön.* (MAGDALIS, RHINODES, PANUS. *Steph.*) Seven species?

    T. barbicornis? *Latr.* Black, opaque; the base of the antennæ ferruginous; the clava fuscous; about two lines; whitethorn hedges; not uncommon. *Pl.* 68, *fig.* 8.

Genus [501]. PISSODES. *Germ.* Three species.

    P. Pini, *Fab.* Rufo-piceous, variegated with paler scales and two transverse bands; about four lines; fir trees; Scotland; very rare. *Pl.* 69, *fig.* 1.

Genus [502]. RHINOCYLLUS. *Germ.* One species.

    R. latirostris, *Latr.* Black, variegated with tufts of yellowish pile; antennæ and tarsi pitchy; about four lines; coasts; not common. *Pl.* 69, *fig.* 2.

Genus [503]. RHINOBATUS. *Megerle.* One species.

    R. Carlinæ, *Oliv.* (planus, *Steph.*) Black, tesselated with cinereous pubescence; about four lines; coasts. *Pl.* 69, *fig.* 3.

Genus [504]. LARINUS. *Schüp.* One species.

    L. Stephensii, *Shuck.* (sturnus? *Steph.*) Black; sides of the thorax ashy; elytra with tufts of grey pile; the third interstice with an interrupted and whitish band; six lines; hedges; Merton, near Kingston. ? British. *Pl.* 69, *fig.* 4.

Genus [505]. LIXUS. *Fab.* Four species.

    L. paraplecticus, *Lin.* Black, densely covered with greenish yellow scales, or decumbent down, which is frequently more or less denuded; antennæ with the base reddish; about six or seven lines; common on the banks of the Thames. *Pl.* 69, *fig.* 5.

BAND II. BRACHYRHYNCHI. *Schön.*

Family [88]. OTIORHYNCHIDÆ. Shuck.

Genus [506]. OTIORHYNCHUS. *Germ.* Nineteen species?

    O. sulcatus, *Fab.* Black, mottled with brown scales; about five lines; common in hedges. *Pl.* 69, *fig.* 6.

Family [89]. OMIADÆ. Shuck.

Genus [507]. OMIAS. *Schön.* (BRACHYSOMUS and partly OTIORHYNCHUS, *Steph.*) Four species.

O. hirsutulus, *Fab.* Blackish brown or testaceous; antennæ and legs testaceous; thorax and elytra covered with erect setæ; nearly two lines; sandy places. *Pl. 70, fig.* 1.

Genus [508]. TRACHYPHLÆUS. *Germ.* Six species.
 T. scabriculus, *Lin.* Entirely of a dirty brown; legs and antennæ slightly testaceous; two to three lines; sandy heaths, in pits. *Pl.* 70, *fig.* 2.

Family [90]. PHYLLOBIIDÆ. Shuck.

Genus [509]. PHYLLOBIUS. *Schön.* Nine species? N. B. All the femora are either toothed or simple.
 P. Pyri, *Lin.* Black, covered with bluish or greenish scales; legs and antennæ red; three to four lines; common upon nettles. *Pl.* 70, *fig.* 3.
Genus [—]. NEMOICUS. *Dillw.* One species.
 N. oblongus, *Lin.* Usually brown, with the head and thorax darker; legs and antennæ testaceous; about three lines; common in hedges. *Pl.* 70, *fig.* 4.

Family [91]. MOLYTIDÆ. Shuck.

Genus [510]. PROCAS. *Steph.* One species.
 P. picipes, *Marsh.* Black, mottled with a few ashy hairs; antennæ and tarsi pitchy; three to four lines; marshy places. *Pl.* 70, *fig.* 5.
Genus [511]. PHYTONOMUS. *Schön.* (HYPERA. *Steph.*) Twenty-two species?
 P. Polygoni, *Lin.* Black or brown, covered with brown scales; the thorax with three longitudinal pale lines, and the elytra with others, of which there are several near the suture, the apex of which has a denticulated one; nearly four lines; sandy corn fields. *Pl.* 70, *fig.* 6.
Genus [512]. PLINTHUS. *Germ.* One species.
 P. caliginosus, *Fab.* Entirely of a dull opaque black; about five lines; chalky districts, under stones. *Pl.* 70, *fig.* 7.
Genus [——]. LEIOSOMA. *Kirby.* One species.
 L. ovatula, *Clairv.* Entirely bright shining black; antennæ and legs occasionally pitchy; about two lines; moist meadows, by sweeping. *Pl.* 70, *fig.* 8.
Genus [513]. MOLYTES. *Schön.* Two species.
 M. Germanus, *Lin.* Intensely black and shining, slightly mottled with ashy hairs; about eight lines; chalky districts. *Pl.* 71, *fig.* 1.
Genus [514]. HYLOBIUS. *Germ.* Two species.
 H. Abietis, *Lin.* Black, covered with brown scales, with

some transverse irregular abbreviated paler bands; six
to eight lines; fir plantations. *Pl.* 71, *fig.* 2.

Genus [515]. TANYSPHYRUS. *Germ.* One species.

    T. Lemnæ, *Fab.* Black, with symmetrical patches of ashy
hair; about one line; banks of weedy ditches. *Pl.* 71,
fig. 3.

Family [92]. CLEONIDÆ. Shuck.

Genus [516]. BABYNOTUS. *Germ.* Three species?

    B. Mercurialis, *Fab.* Black, clothed with dirty brown
scales; three and a half lines; chalky districts. *Pl.* 71,
fig. 5.

Genus [—]. MERIONUS. *Megerle.* Two species.

    M. obscurus, *Fab.* Obscure black, thickly mottled with
fuscous scales; about four and a half lines; sandy heaths,
in pits. *Pl.* 71, *fig.* 4.

Genus [517]. LIOPHLÆUS. *Germ.* One species.

    L. nubilus, *Fab.* Black, covered with ashy scales; elytra
mottled with fuscous; antennæ pitchy—sometimes com-
pletely denuded, and then obscure black; about four to
five lines; hedges, by beating. *Pl.* 71, *fig.* 6.

Genus [518]. ALOPHUS. *Schön.* One species.

    A. triguttatus, *Fab.* Black, densely clothed with fuscous
scales; the elytra with a spot in their centre towards the
base, and one common to both, V shaped, towards the
apex; antennæ pitchy; about four lines; moist meadows,
by sweeping. *Pl.* 71, *fig.* 7.

Genus [519]. GRONOPS. *Schön.* One species.

    G. lunatus, *Fab.* Completely covered with fuscous scales;
the head and two transverse bands on the elytra paler;
about three lines; pits in sandy places. *Pl.* 71, *fig.* 8.

Genus [520]. BOTHYNODERES. *Schön.* One species.

    B. albidus, *Fab.* Blackish brown, variously but symmetri-
cally mottled and maculated with white scales; about
six lines; gravel pits; very rare. *Pl.* 72, *fig.* 1.

Genus [521]. CLEONUS. *Schön.* Four species.

    C. nebulosus, *Lin.* Black, variegated with reddish and
ashy scales; the suture reddish; and the elytra with two
oblique denuded bands; five to nine lines; heaths; New
Forest. *Pl.* 72, *fig.* 2.

Family [93]. BRACHYDERIDÆ. Shuck.

Genus [522]. POLYDROSUS. *Schön.* Twelve species.

    P. undatus, *Fab.* Black, covered with brownish metallic
scales; elytra with two or three undulated bands of

darker scales; antennæ and legs testaceous; two and a
half lines; hedges; not uncommon. *Pl. 72, fig.* 3.

Genus [523]. SITONA. *Germ.* Thirteen species?

S. puncticollis, *Kirby.* Clothed with fuscous scales; the
thorax with two pale lateral stripes; base of antennæ,
tibiæ and tarsi red; about three lines. *Pl. 72, fig.* 4.

Genus [524]. TANYMECUS. *Germ.* One species.

T. palliatus, *Fab.* Closely covered with fuscous scales; the
sides paler; about five lines; sandy places; not common. *Pl. 72, fig.* 5.

Genus [525]. SCIAPHILUS. *Schön.* One species.

S. muricatus, *Fab.* Clothed with ashy or silvery fuscous
scales intermixed with hairs; antennæ and legs pitchy;
about three lines; weedy places in woods. *Pl. 72, fig.* 6.

Genus [526]. STROPHOSOMUS. *Billb.* Eleven species?

S. Coryli, *Fab.* Covered with fuscous and ashy scales;
the apex of the elytra mottled; suture black about one-
third of its length; antennæ and legs testaceous red;
about three lines; on the hazel; common. *Pl. 72, fig.* 7.

Genus [527]. CNEORHINUS. *Schön.* (PHILOPEDON. *Steph.*)
Three species.

C. geminatus, *Fab.* Densely covered with fuscous scales;
the alternate interstices of the elytra paler; two to four
lines; sandy coasts. *Pl. 72, fig.* 8.

SUBDIVISION II. XYLOPHAGI. LATR.

TRIBE I. CYLINDRICI. SHUCK.

Family [94]. HYLESINIDÆ. Shuck.

Genus [528]. HYLASTES. *Erichs.* Nine species?

H. ater, *Fab.* Brownish black, with a few dispersed hairs;
antennæ and legs pitchy; two to two and a half lines;
stumps of old trees. *Pl. 73, fig.* 1.

Genus [529]. DENDROCTONUS. *Erichs.* One species.

D. piniperda, *Lin.* Black and shining; antennæ and tarsi
reddish testaceous; two to two and a half lines; old
fir trees. *Pl. 73, fig.* 2.

Genus [530]. HYLESINUS. *Fab.* Eight species.

H. crenatus, *Fab.* Pitchy black; antennæ and tarsi pitchy
red; two to three lines; stumps of old trees. *Pl. 73,
fig.* 5.

Genus [531]. POLYGRAPHUS. *Erichs.* One species.

P. pubescens, *Fab.* Pitchy brown; the thorax rather
darker; antennæ and legs reddish testaceous; about one
line; decayed trees. *Pl. 73, fig.* 6.

Genus [532]. SCOLYTUS. *Geoffr.* Six species.

 S. Destructor, *Oliv.* Black and shining ; elytra pitchy red or red ; antennæ and legs reddish testaceous; two to three lines; old elms. *Pl. 73, fig.* 3.

Family [95]. TOMICIDÆ. Shuck.

Genus [533]. TRYPODENDRON. *Steph.* (XYLOTERUS. *Erichs.*) One species.

 T. domesticum, *Lin.* Black; the elytra livid, with the suture and sides rather darker; antennæ and legs fuscous; two lines; decayed trees. *Pl. 73, fig.* 5.

Genus [534]. TOMICUS. *Latr.* Twelve species.

 T. Typographus, *Lin.* Testaceous, subpubescent, sometimes pitchy ; three to four lines ; decayed firs. *Pl. 73, fig.* 7.

Family [96]. PLATYPODIDÆ. Shuck.

Genus [535.] PLATYPUS. *Herbst.* Two species ?

 P. cylindrus, *Herbst.* Pitchy black or pitchy red, slightly shining ; antennæ and tarsi reddish testaceous ; three and a half to four lines; decayed oaks. *Pl. 73, fig.* 8.

TRIBE II. DEPRESSI. SHUCK.

Family [97]. CUCUJIDÆ. Steph.

Genus [536]. CUCUJUS. *Fab.* Six species.

 C. Spartii, *Curt.* Pitchy black ; antennæ and legs pitchy red; one and a half line ; beneath the bark of broom. *Pl. 73, fig.* 9.

Genus [537]. ULEIOTA. *Latr.* One species.

 U. planata, *Lin.* Pitchy black ; mouth and legs red ; or entirely testaceous ; two and a half lines ; beneath bark; very rare; ? indigenous. *Pl. 73, fig.* 10.

SUBDIVISION III. LONGICORNES. LATR.

Family [98]. PRIONIDÆ. Leach.

Genus [538]. PRIONUS. *Geoffr.* One species.

 P. coriarius, *Lin.* Pitchy black ; tarsi frequently pitchy red ; twelve to sixteen lines; skirts of and open places in woods and on old trees. *Pl. 74, fig.* 1.

Genus [539]. SPONDYLIS. *Fab.* One species.

 S. buprestoides, *Fab.* Black or pitchy ; about nine lines ; Windsor Forest ; very rare ; ? indigenous. *Pl. 74, fig.* 2.

Family [99]. CERAMBYCIDÆ. Kirby.

Genus [540]. NECYDALIS. *Lin.* Two species.
N. minor, *Lin.* Pitchy; thorax darker; elytra with an oblique pale line upon the disk; three to five lines; Umbelliferæ, near woods. *Pl. 74, fig. 3.*

Genus [541]. AROMIA. *Serville.* One species.
A. Moschata, *Lin.* Blue, green or coppery, shining; antennæ bluish or greenish; ten to seventeen lines; willows, in marshy places. *Pl. 74, fig. 4.*

Genus [542]. CERAMBYX. *Lin.* One species.
C. Cerdo, *Fab.* Black or pitchy, especially towards the apex of the elytra; twelve to sixteen lines; willows, marshy places. *Pl. 78, fig. 1.*

Genus [543]. HYLOTRUPES. *Serville.* One species.
H. Bajulus, *Lin.* Pitchy; the thorax densely clothed with pale grey down, with two denuded spots on the disk; the elytra, with the base and a tranverse band about the middle, also covered with pale pubescence; six to twelve lines; old timber and outhouses. *Pl. 75, fig. 1.*

Genus [544]. CALLIDIUM. *Fab.* Five species.
C. violaceum, *Lin.* Purplish blue or violet; sometimes greenish; four to nine lines; old fir timber, posts and rails. *Pl. 75, fig. 2.*

Genus [545]. ASEMUM. *Esch.* One species.
A. striatum, *Lin.* Pitchy black; thorax slightly pubescent; seven to nine lines; old trees; Scotland. *Pl. 75, fig. 3.*

Genus [546]. GRACILIA. *Serville.* One species.
G. minuta, *Fab.* Reddish brown; thorax redder; antennæ and legs fuscous; two and a half to three lines; old dead elm enclosures. *Pl. 75, fig. 4.*

Genus [547.] CLYTUS. *Fab.* Four species.
C. Arietis, *Lin.* Deep black; the thorax in front and behind, scutellum, an abbreviated transverse mark at the shoulder, an oblique diverging tranverse line in the centre, another directly transverse at two-thirds their length, the apex of the elytra, and the pygidium, all bright brimstone yellow; antennæ at the base and the legs reddish testaceous; six to eight lines; palings and Umbelliferæ; common. *Pl. 75, fig. 6.*

Genus [548]. OBRIUM. *Megerle.* One species.
O. Cantharinum, *Lin.* Bright chesnut yellow, shining; antennæ and legs pitchy; four to five lines; old timber. *Pl. 75, fig. 5.*

K

Family [100]. LAMIIDÆ. Shuck.

Genus [549]. ASTYNOMUS. *Dej.* (ACANTHOCINUS.) One species. N.B. In the female the antennæ are not much longer than the body.

    A. Ædilis, *Lin.* Ashy brown, with a grey down; thorax with four tubercles covered with yellowish pubescence; elytra with two waved brownish bands; eight to ten lines; timber; rare. *Pl. 76, fig. 2.*

Genus [550]. AGAPANTHIA. *Serville.* One species.

    A. Cardui, *Fab.* Black, densely covered with yellowish or pale ashy pubescence; thorax with three more thickly clothed lines; antennæ with the basal joint and apex of the others black; five to eight lines; thistles, in fens. *Pl. 76, fig. 4.*

Genus [551]. APHELOCNEMIA. *Steph.* (MESOSA. *Megerle.*) One species.

    A. nubila, *Oliv.* Black, tesselated and variegated with reddish brown and ashy; elytra with a waved pale mottled band bordered with black; antennæ and tibiæ ringed with white; six to eight lines; decaying oak branches. *Pl. 76, fig. 3.*

Genus [552]. SAPERDA. *Fab.* Eight species.

    S. carcharias, *Lin.* Black, densely clothed with ashy or yellowish pubescence, and sprinkled with denuded punctures; the apex of the joints of the antennæ black; ten to fourteen lines; poplars; Cambridgeshire. *Pl. 77, fig. 1.*

Genus [553]. POGONOCERUS. *Megerle.* Three species.

    P. pilosus, *Fab.* Reddish brown; the base of the elytra with a broad cream-coloured band, beyond which, near the suture, there are three black tufts; antennæ and tibiæ ringed with white; three to four lines; hedges by beating. *Pl. 77, fig. 2.*

Genus [554]. LEIOPUS. *Serville.* One species.

    L. nebulosus, *Lin.* Brown, mottled with ashy pubescence; elytra with a basal and medial dark waved band; three to four and a half lines; dead branches of oak. *Pl. 77, fig. 3.*

Genus [555]. TETROPS. *Kirby.* One species.

    T. præusta, *Lin.* Black, pubescent; with the elytra, except their apex, the anterior legs entirely, and the tibiæ of the remainder, testaceous; two to three lines; hawthorn hedges; common. *Pl. 76, fig. 1.*

Genus [556]. MONOCHAMUS. *Megerle.* Two species.

    M. Sutor, *Lin.* Black, with scattered ashy pubescent

spots; scutellum white, with a central denuded line; eleven to thirteen lines; ash trees and willows. *Pl. 77, fig. 4.*

Genus [557]. LAMIA. *Fab.* One species.

    L. textor, *Lin.* Black, mottled with a few scattered pale pubescent spots; seven to fourteen lines; decaying willows. *Pl. 78, fig. 2.*

Family [101]. LEPTURIDÆ. Leach.

Genus [558]. RHAGIUM. *Fab.* Three species.

    R. Inquisitor, *Lin.* Black, mottled with yellowish hair; the elytra with two testaceous irregular transverse bands, with a black smooth lateral spot between them; seven to eleven lines; old wood, ash trees. *Pl. 78, fig. 3.*

Genus [559]. TOXOTUS. *Megerle.* One species.

    T. meridianus, *Lin.* Either testaceous with the head, thorax, knees and tarsi black; or entirely black; or black; with the legs reddish testaceous, except the knees; six to fourteen lines; common on ash trees. *Pl. 78, fig. 4.*

Genus [560]. PACHYTA. *Megerle.* Three species.

    P. octomaculata, *Fab.* Black; the elytra livid, with eight black spots, two placed laterally, one apical, and one on the disk near the scutellum; four to six lines; Umbelliferæ in woods. *Pl. 79, fig. 4.*

Genus [561]. GRAMMOPTERA. *Serville.* Five species.

    G. præusta, *Fab.* Black, clothed with close decumbent golden down, excepting the head and the apex of the elytra; the base of the antennæ and the legs testaceous; four to five lines; Umbelliferæ in the New Forest. *Pl. 79, fig. 3.*

Genus [562]. LEPTURA. *Lin.* Ten species.

    L. quadrifasciata, *Lin.* Deep black; the elytra with four transverse interrupted waved testaceous bands; the apical joints of the antennæ sometimes red; seven to ten lines; flowers, in woods. *Pl. 79, fig. 2.*

Genus [563]. STRANGALIA. *Serville.* Two species.

    S. elongata, *De Geer.* Black; the elytra testaceous, with an angulated transverse band near the base, frequently variously interrupted, another broader one towards the apex, with a large lateral spot between these and the extreme apex of the elytra also black; the four anterior legs testaceous, their tarsi black; the posterior pair either entirely black, or the base of their tibiæ only testaceous; five to eight lines; common on flowers, in woods. *Pl. 79, fig. 1.*

Subdivision IV. EUPODA. Latr.

Family [102]. Crioceridæ. Leach.

Genus [564]. Donacia. *Fab.* Twenty-one species.
D. Menyanthidis, *Fab.* Brassy green; antennæ and legs reddish testaceous; about six lines; aquatic plants. *Pl.* 80, *fig.* 1.

Genus [565]. Macroplea. *Hoffmsg.* One species.
M. Zostèræ, *Fab.* Livid, with the base of the antennæ, two spots on the thorax, and some narrow longitudinal striæ on the elytra, black; about three lines; on Zostera marina, near Hull, &c. *Pl.* 80, *fig.* 2.

Genus [566]. Crioceris. *Geoffr.* Seven species.
C. merdigera, *Lin.* Black, with the thorax and elytra reddish testaceous; about four lines; on the white lily. *Pl.* 80, *fig.* 3.

Genus [567]. Zeugophora. *Kungi.* Two species.
Z. subspinosa, *Fab.* Reddish testaceous; elytra blue-black; about two lines; common on the aspen. *Pl.* 80, *fig.* 4.

Genus [568]. Orsodacna. *Latr.* Three species?
O. nigriceps, *Latr.* Variable in colour, but usually testaceous, with a spot upon the thorax, and a narrow line at the suture, black; or entirely black; about three lines; herbage, in damp woods. *Pl.* 80, *fig.* 5.

Genus [569]. Psammæchus. *Bond.* One species.
P. bipunctatus, *Fab.* Testaceous, with the head, apical joints of the antennæ, and two spots on the disk of the elytra, beyond the middle, black; one and a half line; herbage, in marshy places. *Pl.* 80, *fig.* 6.

Subdivision V. CYCLICA. Latr.

Tribe I. CASSIDARIÆ. Latr.

Family [103]. Hispidæ. Kirby.

Genus [570]. Hispa. *Lin.* One species.
H. atra, *Lin.* Entirely of an opaque black; about one line; on nettles; very rare; ? indigenous. *Pl.* 84, *fig.* 5.

Family [104]. Cassididæ. Leach.

Genus [571]. Cassida. *Lin.* Eighteen species.
C. equestris, *Fab.* Pale green, with the margins of the elytra sometimes discoloured; antennæ and legs testa-

ceous ; the apex of the latter obscure ; about four lines ;
common on the burdock. *Pl.* 84, *fig.* 6.

Family [105]. GALERUCIDÆ. Steph.

Genus [572]. AUCHENIA. *Marsh.* One species.
    A. quadrimaculata, *Lin.* Reddish testaceous, with the head
    and four spots on the elytra, two basal and two apical,
    black ; apex of antennæ obscure ; about four lines ;
    marshy meadows. *Pl.* 81, *fig.* 1.
Genus [573]. ADIMONIA. *Schr.* Two species.
    A. Halensis, *Lin.* Testaceous, with the antennæ, legs and
    two spots on the thorax obscure ; the crown of the head
    and the elytra of a bright green ; three to four lines ;
    hedges ; not uncommon. *Pl.* 81, *fig.* 2.
Genus [574]. GALERUCA. *Geoffr.* Eleven species.
    G. Tanaceti, *Lin.* Entirely black, slightly shining ; five to
    six lines ; meadows and on plants. *Pl.* 81, *fig.* 3.
Genus [575]. LUPERUS. *Geoffr.* Two species.
    L. flavipes, *Lin.* Bright shining black and smooth : thorax,
    legs and base of antennæ yellow ; about two lines ; in
    hazel hedges ; common. *Pl.* 81, *fig.* 5.
Genus [576]. CALOMICRUS. *Dillwyn.* One species.
    C. circumfusus, *Marsh.* Bright shining black ; base of
    antennæ, thorax in front and the elytra externally pale
    testaceous ; about one and a half line ; plants and herb-
    age, and birchwood ; common. *Pl.* 81, *fig.* 4.

Family [106]. HALTICIDÆ. Kirby.

Genus [577]. HALTICA. *Illig.* Forty-three species ?
    H. Nemorum, *Lin.* Shining black, with a broad pale livid
    streak down the elytra ; bare of antennæ, tibiæ and tarsi
    testaceous ; about a line ; on culinary plants ; common.
    Pl. 82, *fig.* 1.
Genus [578]. THYAMIS. *Steph.* Thirty-eight species?
    T. femoralis, *Marsh.* Testaceous, with the head, apex of
    the antennæ, suture of the elytra, and the posterior legs,
    black ; about one and a half line ; hedges, by sweeping.
    Pl. 82, *fig.* 2.
Genus [579]. MANTURA. *Steph.* Six species.
    M. semiænea, *Fab.* Greenish black, with the base of the
    antennæ, anterior legs, and posterior tibiæ and tarsi, and
    the external portion of the elytra, reddish testaceous ;
    about one and a half line ; hedges. *Pl.* 82, *fig.* 3.

Genus [580]. MACROCNEMA. *Megerle.* Seventeen species?
    M. Hyoscyami, *Lin.* Bluish black; base of antennæ, four anterior legs, and the posterior tibiæ and tarsi reddish testaceous; one and a half line; plants and herbage. *Pl.* 82, *fig.* 4.

Genus [581]. CARDIAPUS. *Curt.* One species.
    C. Matthewsii, *Curt.* Greenish black; base of the antennæ, and the tibiæ and tarsi testaceous; about one line; herbage; not common. *Pl.* 82, *fig.* 6.

Genus [582]. DIBOLIA. *Latr.* Two species.
    D. Cynoglossi, *Ent. Heft.* Brassy green; base of antennæ, four anterior legs, posterior tibiæ and tarsi reddish testaceous; one and a half line; on the hound's tongue, in wastes. *Pl.* 82, *fig.* 5.

Genus [583]. CHÆTOCNEMA. *Steph.* Six species.
    C. concinna, *Marsh.* Brassy green; base of the antennæ, the tibiæ and tarsi red; about one line. *Pl.* 83, *fig.* 1.

Genus [584]. SPHÆRODERMA. *Steph.* Five species.
    S. testaceum, *Fab.* Entirely of a reddish testaceous; one and a half line; on thistles; common. *Pl.* 82, *fig.* 2.

Genus [585]. MNIOPHILA. *Steph.* One species.
    M. Muscorum, *Müller.* Brassy black; antennæ and legs testaceous; half line; in moss; not common, *Pl.* 81, *fig.* 6.

TRIBE III. CHRYSOMELINÆ. LATR.

Family [107]. CHRYSOMELIDÆ. Leach.

Genus [586]. TIMARCHA. *Megerle.* Two species.
    T. lævigata, *Lin.* Entirely of a bright blue black; seven to nine lines; common amongst rank herbage and on heaths. *Pl.* 83, *fig.* 6.

Genus [587]. MELASOMA. *Dillwyn.* Four species.
    M. Populi, *Lin.* Brassy green, with the elytra reddish testaceous; five to six lines; on sapling poplars in woods. *Pl.* 84, *fig.* 1.

Genus [588]. CHRYSOMELA. *Lin.* Twenty-six species?
    C. fulgida, *Lin.* Bright golden green, with occasionally a couple of coppery red streaks; four to five lines; marshy places. *Pl.* 83, *fig.* 5.

Genus [589]. PHÆDON. *Megerle.* Twelve species?
    P. fastuosum, *Lin.* Bright coppery green, with alternate bright blue stripes; about three lines; hedges and dry ditches. *Pl.* 83, *fig.* 3.

Genus [590]. PRASOCURIS. *Latr.* (HELODES. *Steph.* Two species.

P. Beccabungæ, *Payk.* Greenish or bluish; antennæ and legs black; about two lines; weeds, in marshy places. *Pl.* 83, *fig.* 4.

Family [108]. CRYPTOCEPHALIDÆ. Kirby.

Genus [591]. CLYTHRA. *Laichartg.* Four species.

C. quadripunctata, *Lin.* Black; elytra bright testaceous, with four black spots, two basal and two medial; about five lines; heaths and woods; common. *Pl.* 84, *fig.* 2.

Genus [592]. CRYPTOCEPHALUS. *Geoffr.* Twenty-two species.

C. Coryli, *Fab.* Black, with the base of the antennæ and elytra deep chesnut red in the male, or thorax and elytra deep chesnut red in the female; three to four lines; on the hazel; not common. *Pl.* 84, *fig.* 3.

Genus [593]. EUMOLPUS. *Fab.* Three species.

E. Dillwynii, *Steph.* Coppery red, shining; elytra with two bronzy medial depressions; tip of antennæ and tarsi fuscous; two and a quarter lines; South Wales; very rare. *Pl.* 84, *fig.* 4.

TRIBE IV. CLAVIPALPI. LATR.

Family [109]. TRITOMIDÆ. Shuck.

Genus [594]. PHALACRUS. *Payk.* Twenty-four species?

P. coruscus, *Payk.* Entirely bright shining black; about one line; on flowers, in meadows; common. *Pl.* 85, *fig.* 1.

Genus [595]. ALEXIA. *Steph.* Three species.

P. pilifera, *Müll.* Reddish testaceous, densely covered with long erect pile; about half a line; beneath bark; New Forest. *Pl.* 85, *fig.* 2.

Genus [596]. TRITOMA. *Fab.* One species.

T. bipustulatum, *Fab.* Black and shining; elytra with a large red spot at the shoulders; base of the antennæ and the tarsi also red; about two lines; beneath bark and in fungi. *Pl.* 85, *fig.* 3.

Genus [597]. TRIPLAX. *Payk.* Five species?

T. russica, *Lin.* Bright chesnut red; elytra shining black; about three lines; common in fungi. *Pl.* 85, *fig.* 4.

Section IV. TRIMERA. Latreille.

Tribe I. FUNGICOLA. Latr.

Family [110]. ENDOMYCHIDÆ. Leach.

Genus [598]. ENDOMYCHUS. *Payk.* One species.

C. coccineus, *Lin.* Bright vermilion red; head, antennæ, a central spot on the thorax, two large discoïdal spots on the elytra placed longitudinally, and the legs black; about two and a half lines; fungi; local. *Pl.* 85, *fig.* 5.

Genus [599]. LYCOPERDINA. *Latr.* One species.

L. Bovistæ, *Payk.* Black or pitchy; antennæ and legs pitchy red; about two lines; in the puff-ball; common. *Pl.* 85, *fig.* 6.

Tribe II. APHIDAPHAGI. Latr.

Family [111]. COCCINELLIDÆ. Leach.

Genus [600]. CHILOCORUS. *Leach.* Four species.

C. quadriverrucatus, *Fab.* Black; each elytron with two obscure red spots, the first humeral and kidney-shaped, the second just beyond the middle and near the suture, irregular; about two lines; on trees; common. *Pl.* 86, *fig.* 1.

Genus [601]. COCCINELLA. *Lin.* Thirty species?

C. ocellata, *Lin.* Black; the thorax with two large lateral and two central marginal white spots; the elytra reddish yellow, and each with seven black spots surrounded with white, placed two, three and two; tarsi fuscous; four to five lines; fir plantations; not uncommon. *Pl.* 86, *fig.* 2.

Genus [602]. SPHÆROSOMA. *Kirby.* One species.

S. Quercus, *Leach.* Pitchy black; antennæ and legs testaceous; about half a line; on oaks; not common. *Pl.* 86, *fig.* 3.

Genus [603]. SCYMNUS. *Herbst.* Fifteen species.

S. bipustulatus, *Thunb.* Black; each elytron with a large lateral testaceous red spot; antennæ and legs testaceous; one and a quarter line; on plants and flowers. *Pl.* 86, *fig.* 4.

Genus [604]. RHYZOBIUS. *Steph.* One species.

R. Litura, *Fab.* Reddish testaceous; the elytra with a curved interrupted marking, placed just beyond the middle; about a line; grass, by sweeping. *Pl.* 85, *fig.* 5.

Genus [605]. CACICULA. *Megerle.* Two species.

C. pectoralis, *Fab.* Shining testaceous red; about one line; grass, by sweeping; common. *Pl.* 86, *fig.* 6.

SUPPLEMENT

Of Genera either taken alive, (having been introduced from abroad,) or reputed to have been taken alive in England, but of which there is no well authenticated proof of their being indigenous.

———

1. Oxystomus, *Dej.*; anglicanus, *Steph.* Deep black; antennæ and palpi pitchy; tarsi reddish; five and a half lines; found at Peckham. *Supp. Pl.* 1, *fig.* 1.
2. Distomus, *Leach*; fulvipes, *Latr.* Pitchy black; thorax reddish, as are also the antennæ and legs; five lines; said to occur in Devonshire. *Supp. Pl.* 1, *fig.* 2.
3. Alpæus castaneus, *Bonelli.* Pitchy; antennæ and legs red; four and a half lines; said to occur in Devonshire. *Supp. Pl.* 1, *fig.* 3.
4. Procrustes, *Bonelli*; coriaceus, *Lin.* Entirely deep black; eighteen lines; said to occur near Portsmouth. *Supp. Pl.* 1, *fig.* 4.
5. Sogines, *Leach*; punctulatus, *Illig.* Entirely opaque black; six lines; said to occur in Devonshire. *Supp. Pl.* 2, *fig.* 1.
6. Cophosus. *Dej.*; elongatus, *Sam.* Shining black; tarsi reddish; seven and a half lines; said to occur in Devonshire. *Supp. Pl.* 2, *fig.* 2.
7. Cheporus, *Latr.*; metallicus, *Fab.* Of a brilliant brass or coppery colour; antennæ and legs pitchy black; six lines; said to have been found in Cambridgeshire and Kent. *Supp. Pl.* 2, *fig.* 3.
8. Masticus palpalis, *Latr.* Opaque black; antennæ and legs pitchy; about two lines. *Supp. Pl.* 2, *fig.* 4.
9. Tribolium castaneum, *Macleay.* Entirely of an opaque reddish chesnut; one line and three quarters; boxes of Chinese insects. *Supp. Pl.* 2, *fig.* 5.
10. Oryctes, *Illig.*; nasicornis, *Lin.* Bright chesnut brown;

head and thorax slightly darker ; sixteen lines ; said to
have been found at Chelsea. *Supp. Pl.* 4, *fig.* 1.

11. DYNASTES, *Macleay;* inermis, *Martin.* ? Juvencus, *Fab.*
Pitchy black ; antennæ and legs pitchy red : six and a
half lines ; said to have been found among the rejectamenta of a flood. *Supp. Pl.* 4, *fig.* 2.

12. VALGUS, *Scriba ;* hemipterus, *Lin.* Pitchy black, mottled
with ashy white scales ; the pygidium white, with two
denuded transverse spots ; four and a half lines ; said to
have occurred in the vicinity of London. *Supp. Pl.* 4,
fig. 3.

13. CHRYSOBOTHRIS, *Esch. ;* chrysostigma, *Lin.* Bronzy or
coppery ; the thorax coppery red ; the elytra with two
golden depressions on the disk ; antennæ and legs coppery ; six lines ; said to have been taken at large. *Supp.
Pl.* 3, *fig.* 1.

14. LAMPRA, *Megerle ;* rutilans, *Fab.* Brilliant golden green ;
the sides of the thorax and elytra of a rich golden red ;
six lines ; said to have been found in timber in Derbyshire and Kent. *Supp. Pl.* 3, *fig.* 2.

15. DICERCA, *Esch. ;* ænea, *Lin.* Entirely of a reddish bronzy
tint ; nine lines ; said to have been found in Devonshire.
Supp. Pl. 3, *fig.* 3.

16. ANCYLOCHEIRA, *Esch. ;* octoguttata, *Lin.* Deep dark blue,
with the sides of the thorax and each elytron with five
yellow spots, one humeral and four placed longitudinally ;
five and a half lines ; said to have been found at large.
Two others of this genus have also been introduced.
Supp. Pl. 3, *fig.* 4.

17. MELANOPHILA, *Esch. ;* tarda, *Fab.* Entirely of a bright
shining blue ; four and a half lines ; said to have been
found near Windsor. *Supp. Pl.* 3, *fig.* 5.

18. PTOSIMA, *Serville ;* novem-maculata, *Lin.* Blue black,
shining, and having nine yellow spots, one at the vertex,
two transversely on the thorax, and three on each elytron ; five lines ; once taken in abundance at Cocker-
mouth, in Cumberland, by Dr Leach ; doubtlessly imported in timber. *Supp. Pl.* 3, *fig.* 6.

19. DRASTERIUS, *Esch. ;* bimaculatus, *Fab.* Black ; the elytra
obscurely reddish, their apex for about one-third their
length black, enclosing two pale spots and a widely interrupted narrow black band, about the middle ; three
and a half lines ; said to have been found in Devonshire.
Supp. Pl. 5, *fig.* 1.

20. Enicopus, *Steph.*; ater, *Fab.* Entirely of a deep black, slightly shining and densely pubescent; three lines and three-quarters; said to have been found in Devonshire. *Supp. Pl. 5, fig. 2.*

21. Rhyzopertha, *Steph.*; Pusilla, *Fab.* Deep chesnut brown; antennæ reddish testaceous; one and a half line; found occasionally in foreign roots and seeds. *Supp. Pl. 5, fig. 3.*

22. Rhipidius, *Thunb.*; anceps, *Steph.* (? Symbius Blattarum, *Sunderv.*) Pitchy black; elytra brown; legs fuscous; knees and antennæ dirty testaceous; two lines; found by Mr. Stephens amongst paper from Portsmouth. It is very probably *Symbius*, and, if so, parasitical upon the cockroach; and, like the *Evania appendigaster*, is occasionally found on board ships. *Supp. Pl. 5, fig. 4.*

23. Phlæobius, *Schön.*; griseus, *Fab.* Brown black, tesselated with a reddish ashy grey; two and a half lines; it has occurred in several counties, but being a native of New Holland has evidently been imported. *Supp. Pl. 6, fig. 1.*

24. Caryoborus, *Schön.*; cruciger, *Steph.* Grey black, with an ashy white cross upon the base of the elytra; the thorax and remainder of the elytra mottled; anterior legs pale testaceous; and base of posterior thighs white; two and a half lines; found in West India seeds. *Supp. Pl. 6, fig. 2.*

25. Rhytirhinus, *Schön.*; porcatus, *Marsh.* Black, clothed with brown scales; thorax and tibiæ paler; two and a half lines; found on an exotic flower in a garden at Hammersmith, doubtlessly imported from the Cape. *Supp. Pl. 6, fig. 3.*

26. Hypothenemus eruditus, *Westw.* Pitchy; the thorax, antennæ and elytra reddish; about one-third of a line; found in abundance in the cover of an old book. *Supp. Pl. 6, fig. 4.*

27. Purpuricenus, *Ziegler;* Koehleri, *Lin.* Deep opaque black; elytra of a bright red, with a diamond shaped spot in their centre; fifteen to seventeen lines; said to have occurred near London. *Supp. Pl. 7, fig. 1.*

28. Eburia, *Serville;* quadrimaculata, *Fab.* Testaceous, thorax with two black tubercles; elytra with four ivory coloured geminated spots, one at the base, and the second about the middle; eleven lines; found in Essex; doubtlessly imported. *Supp. Pl. 7, fig. 2.*

29. ELAPHIDION, *Serville;* spinicorne, *Fab.* Reddish testaceous, densely covered with short ashy pubescence, with mottled denuded spots ; antennæ and legs very slightly clothed ; ten lines ; found at Bermondsey amongst timber. *Supp. Pl. 7, fig. 3.*

30. TETRAOPES, *Dalmann;* tornator, *Fab.* Pale reddish testaceous; thorax with four black spots ; the elytra with a small black spot at the shoulder, and another on the disk beyond the middle, and an ashy irregular macula at the middle, and another larger at the apex ; antennæ fuscous ; legs black ; six lines; found amongst timber, but a native of North America. *Supp. Pl. 8, fig. 1.*

31. CYLINDERA, *Newm.;* luteus, *Marsh.* (CURTOMERUS, *Steph.*) Entirely testaceous ; in the female (C. pallida, *Newm.*) the antennæ are shorter and not fringed, and the thighs are less robust ; four to five lines ; found amongst timber, and imported. *Supp. Pl. 8, fig. 2.*

32. PENICHROA, *Steph.;* fasciata, *Wilkin.* Entirely dirty testaceous ; elytra with an indistinct band across the middle ; seven to nine lines ; once found in abundance at Norwich, but evidently imported. *Supp. Pl. 8, fig. 3.*

33. ARHOPALUS, *Serville;* fulminans, *Fab.* Black, with an ashy pubescence ; the thorax with three denuded spots ; the elytra with several irregular angulated bands ; about six lines ; taken at Kensington, but doubtlessly imported from North America. *Supp. Pl. 8, fig. 4.*

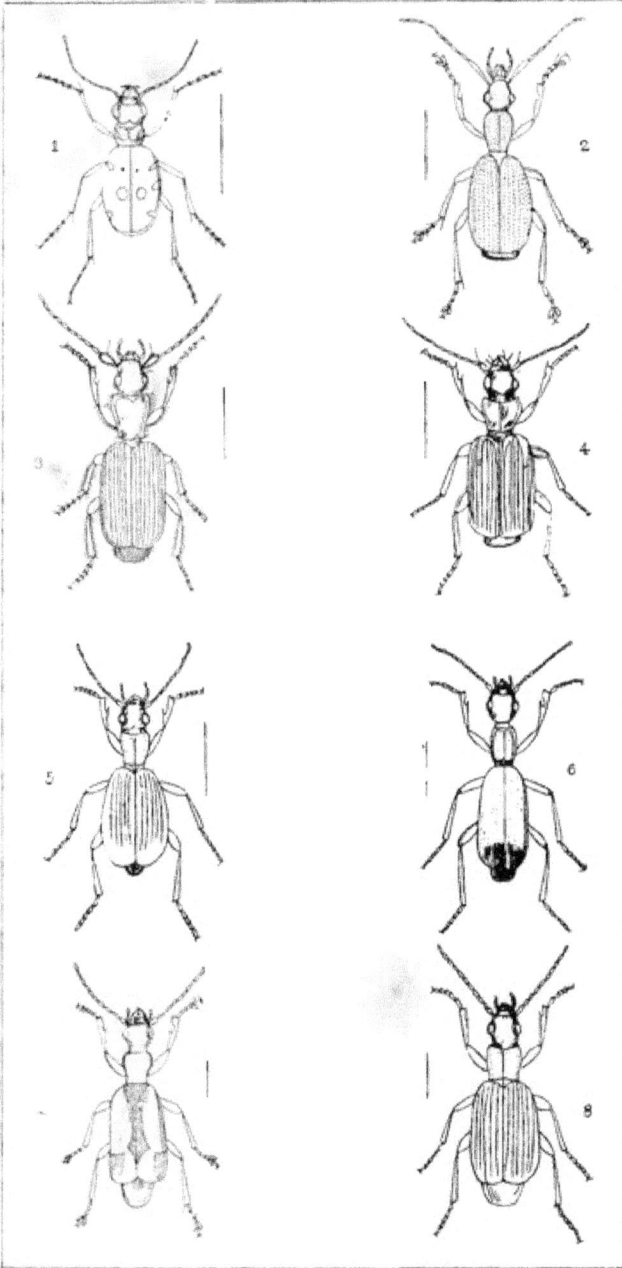

1 CICINDELA
2 DRYPTA
3 POLISTICHUS
4 CYMINDIS

5 BRACHINUS
6 ODACANTHA
7 DEMETRIAS
8 DROMIUS

Pl. 2

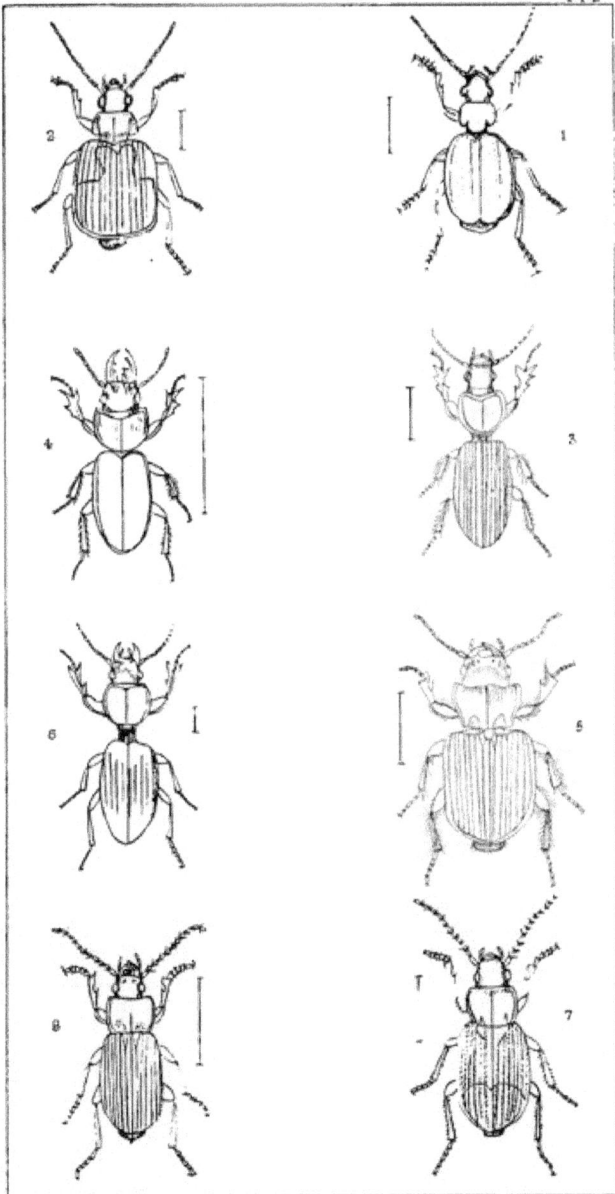

1 LEBIA
2 LAMPRIAS
3 SCARITES
4 CLIVINA

5 DISCHIRIUS
6 SELNOPHORUS
7 ANISODACTYLUS
8 DIACHROMUS

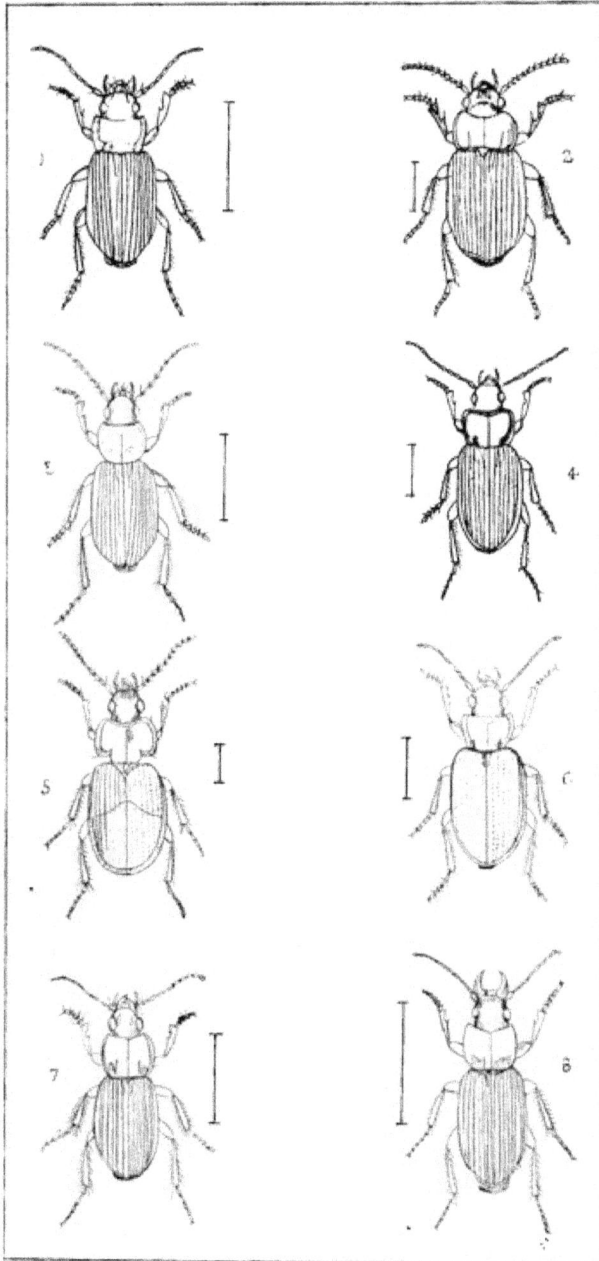

1 HARPALUS
2 ACTEPHILUS
3 OPHONUS
4 STENOLOPHUS

5 MASOREUS.
6 POGONUS
7 PŒCILUS
8 OMASEUS.

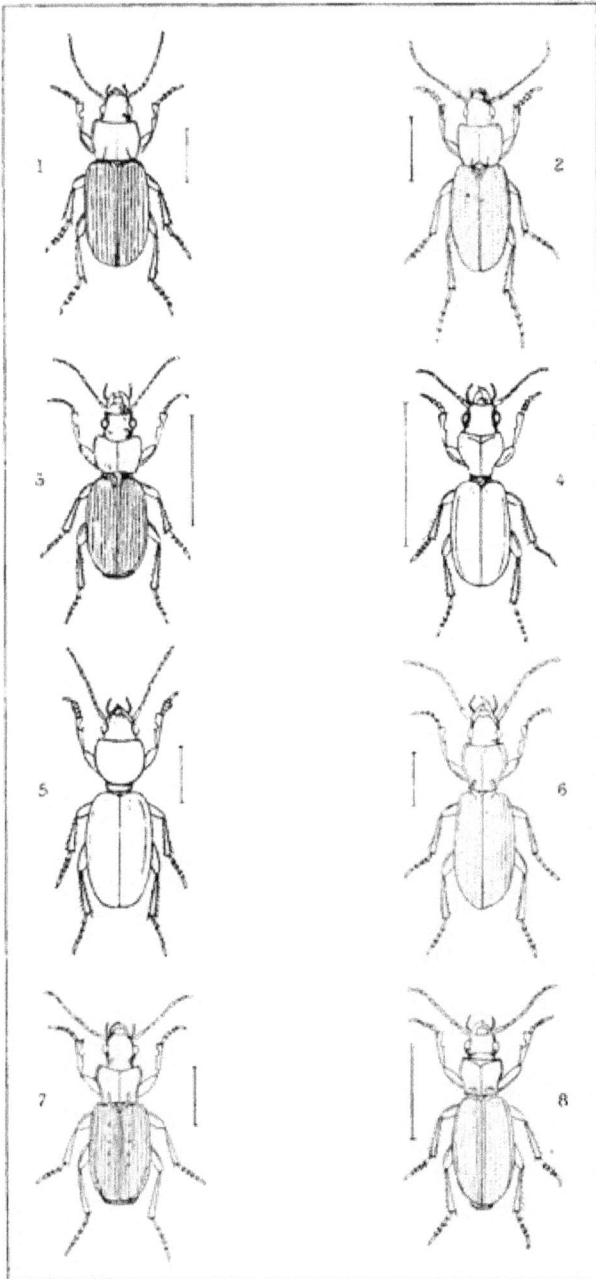

Pl. 4

1. ARGUTOR
2. PLATYDERUS
3. STEROPUS
4. BROSCUS

5. MISCODERA
6. STOMIS
7. PATROBUS
8. PTEROSTICHUS

Pl 5

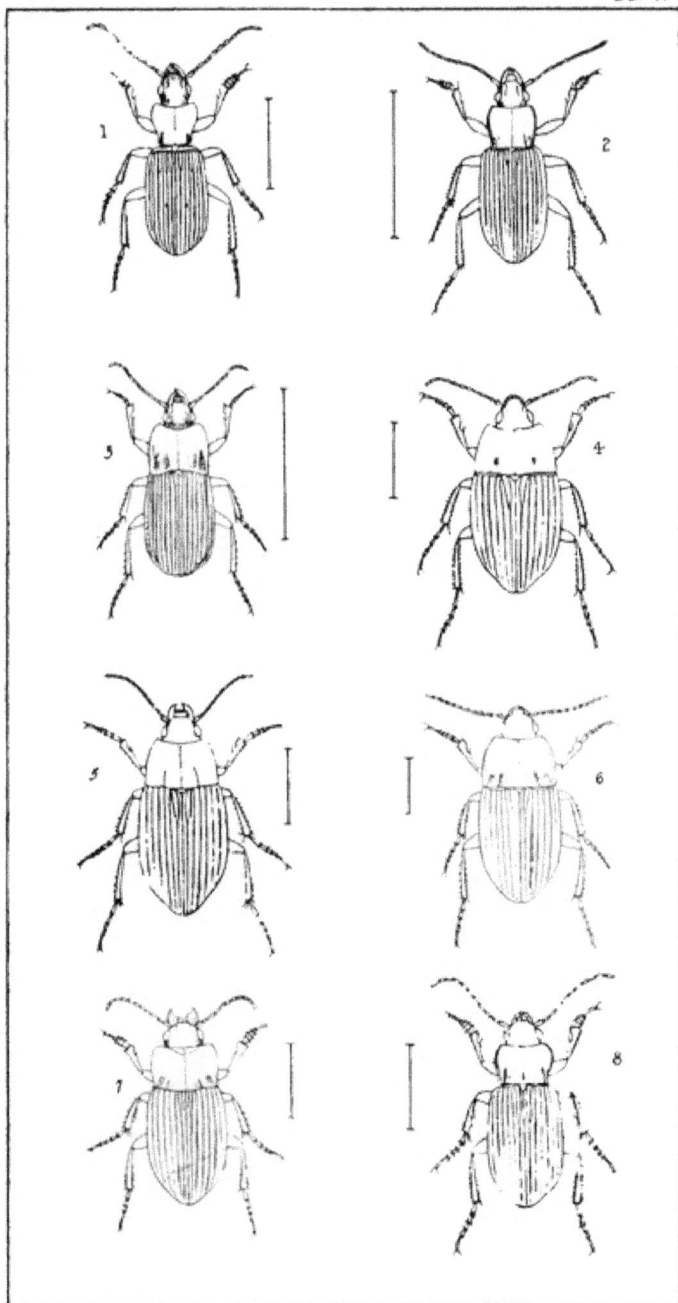

1. ADELOSIA.
2. PLATYSMA.
3. ABAX.
4. AMARA

5. CELIA.
6. ACRODON,
7. BRADYTUS.
8. CURTONOTUS

Pl 6.

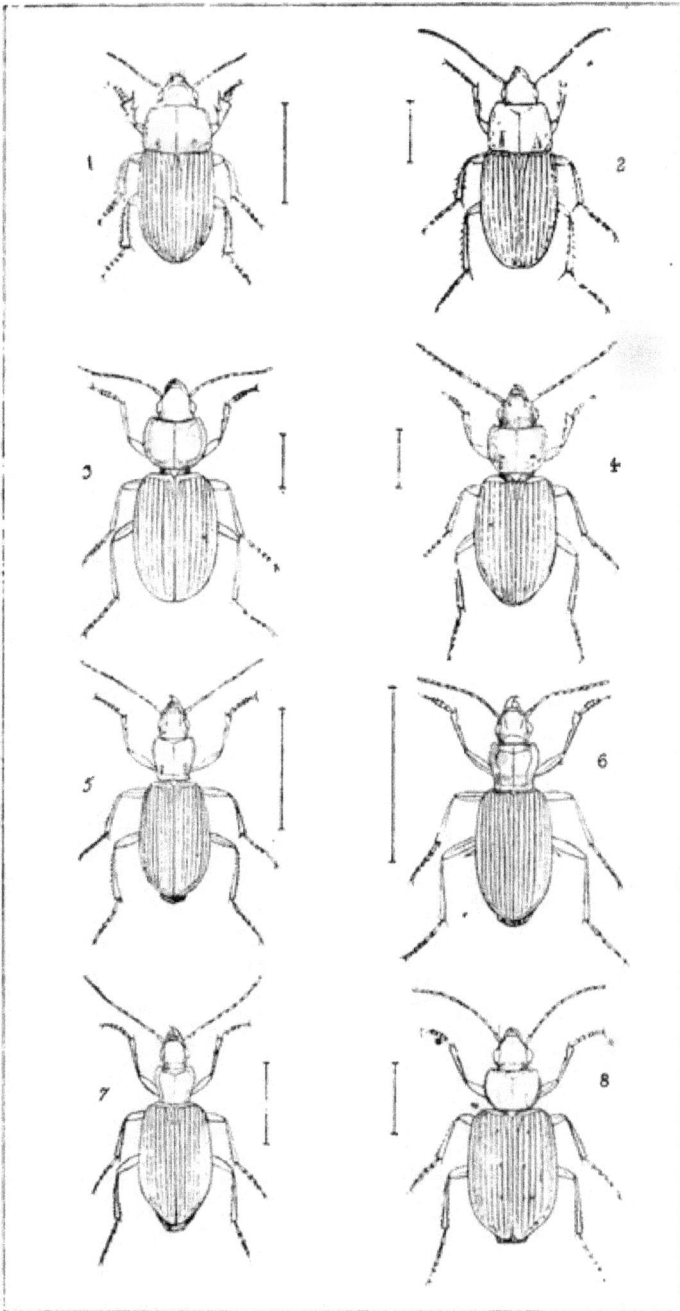

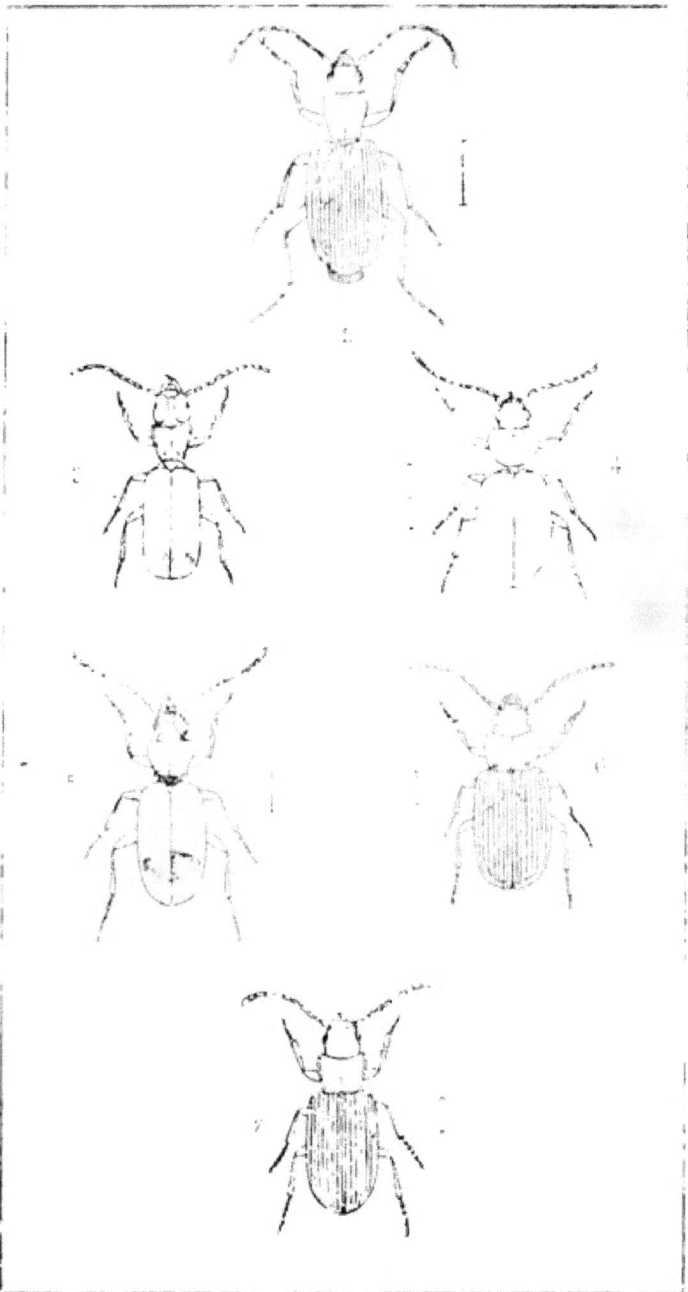

[1. ODONTONYX.]
2. ANCHOMENUS.
3. ÆPUS.

4. EPAPHIUS.
5. BLEMUS.
6. BRADYCELLUS

7. TRECHUS.

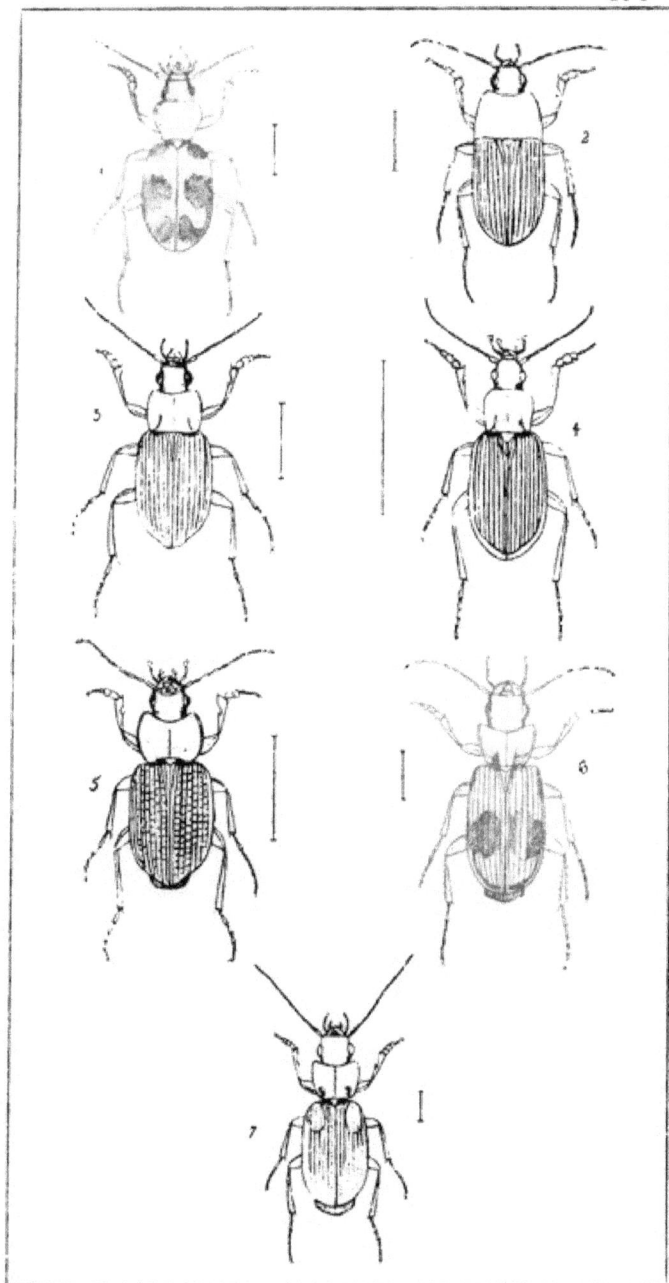

Pl. 8

1. CALLISTUS
2. OODES
3. CHLÆNIUS.

4 EPOMIS
5 LICINUS
6 BADISTER.

7. TRIMORPHUS

Pl 9

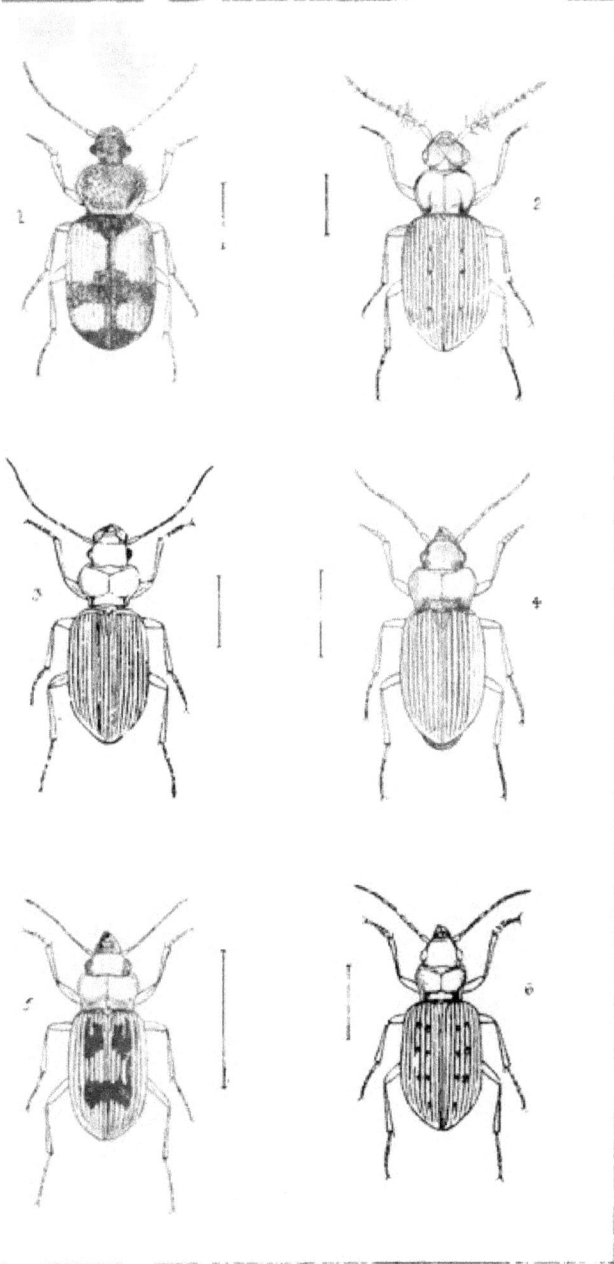

1 PANAGÆUS
2 LORICERA
3 LEISTUS

4 HELOBIA
5 NEBRIA
6 PELOPHILA

Pl 10

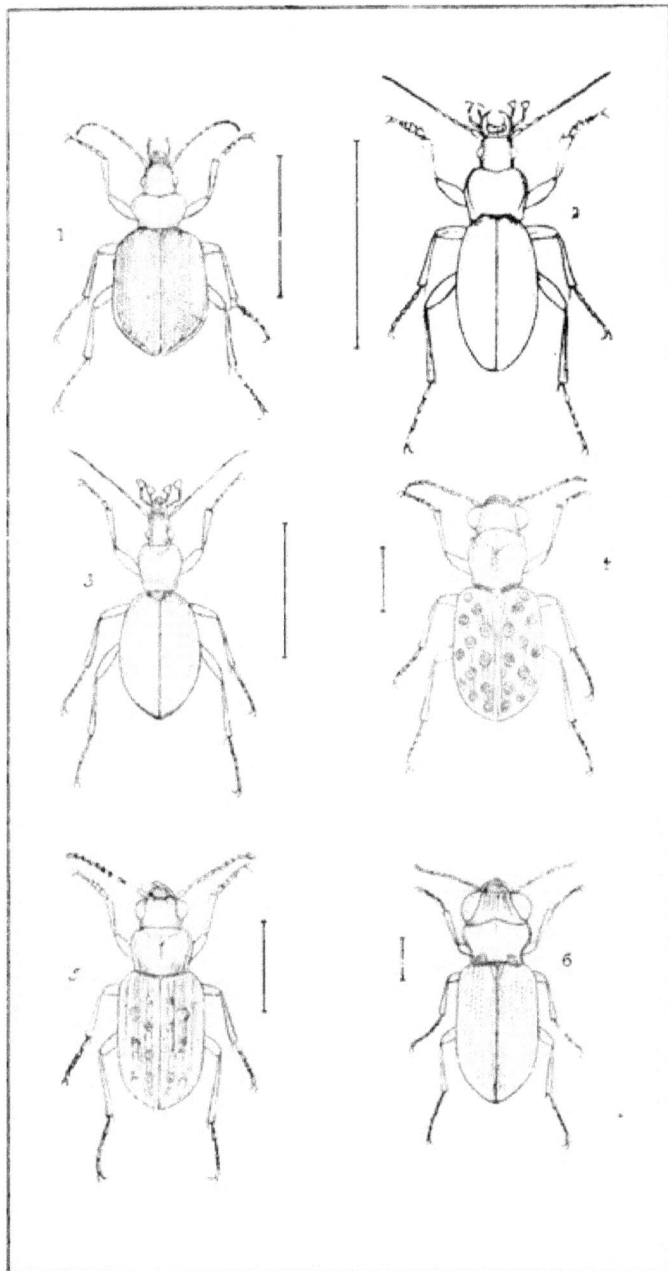

1. CALOSOMA
2. CARABUS
4. ELAPHRUS
5. BLETHISA

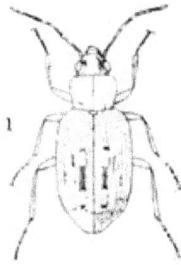

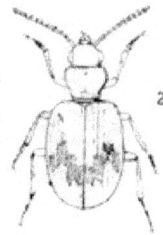

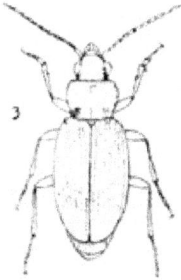

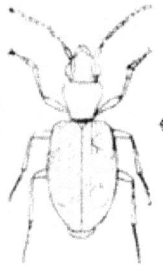

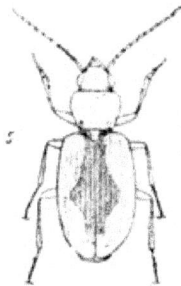

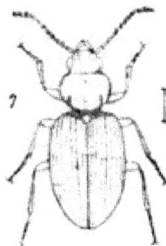

Pl 11

1 BEMBIDIUM

2 TACHYPUS

3 NOTAPHUS

4 LOPHA

5 PERYPHUS

6 OCYS

7 PHILOCHTHUS

Pl. 12

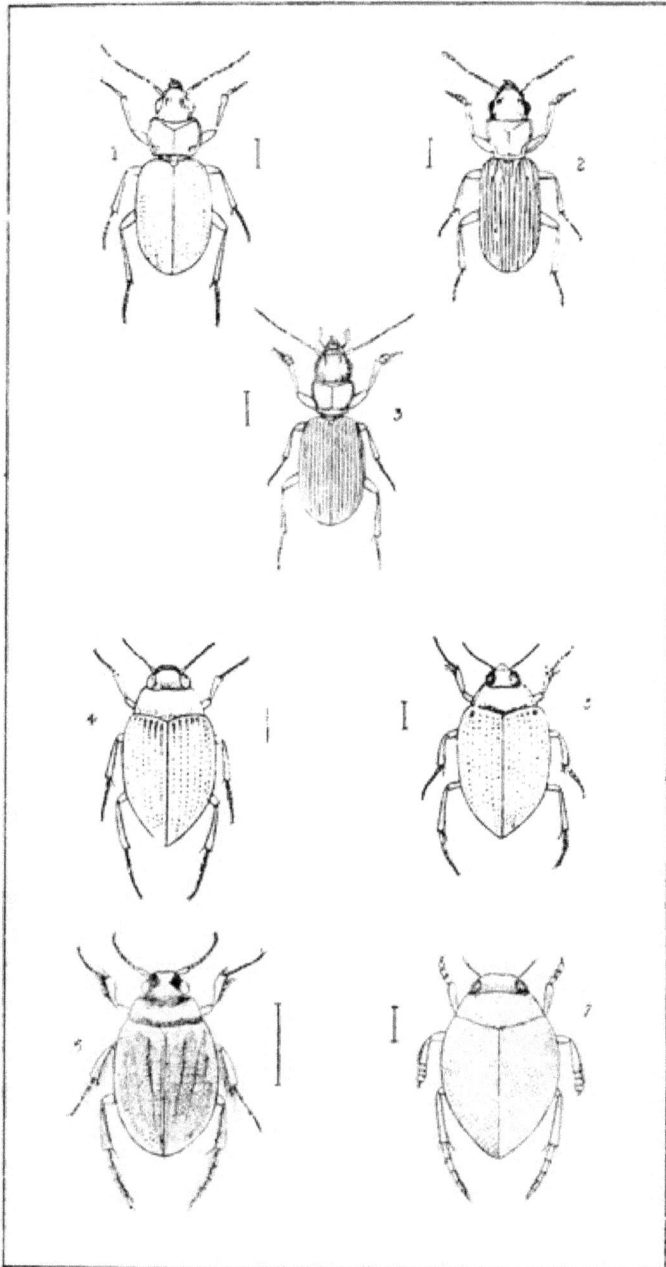

1. TACHYS.
2 CILLENUM.
3 LYMNÆUM

4 HALIPLUS.
5 CNEMIDOTUS
6 PELOBIUS

7 HYPHIDRUS.

Pl 13

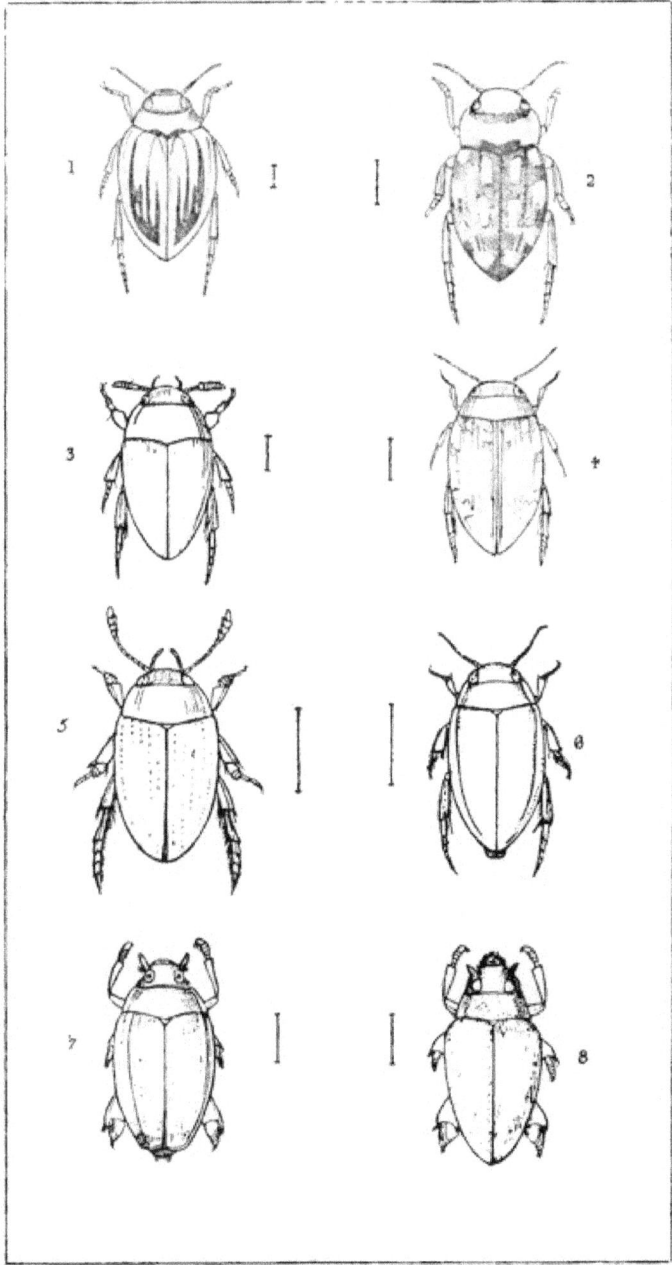

1	HYGROTUS	5	AGABUS
2	HYDROPORUS	6	ILYBIUS
3	NOTERUS	7	GYRINUS
4	LACCOPHILUS	8	ORECTOCHILUS

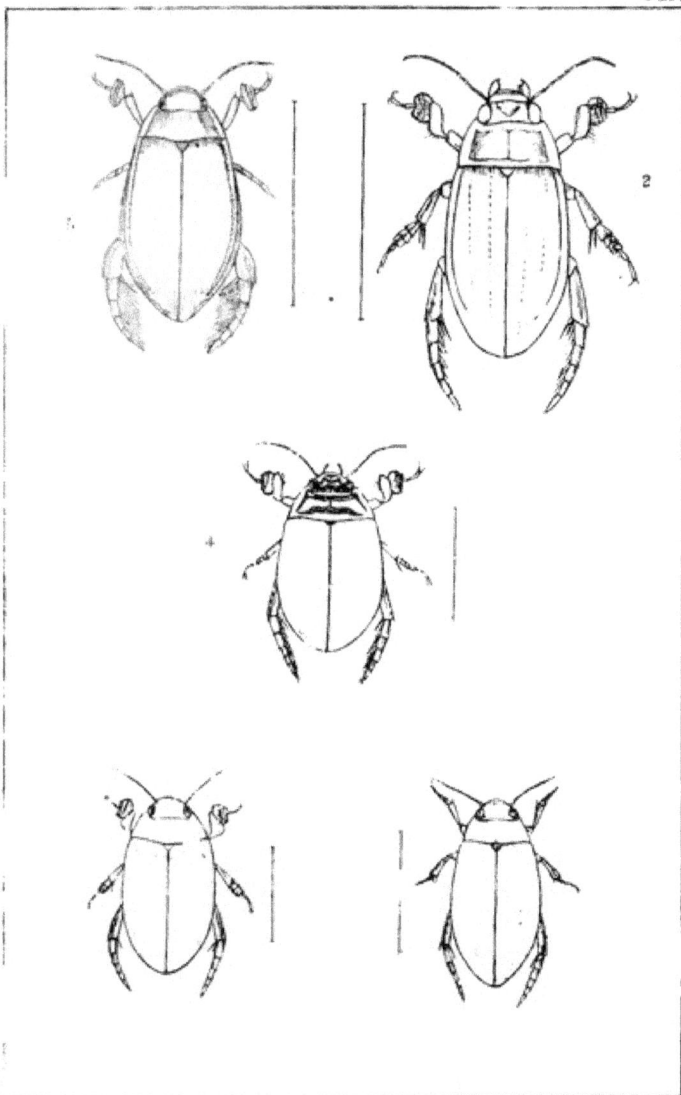

PL1.4

1. COLYMBETES
2. DYTISCUS
3. HYDATICUS.
4. ACILIUS.
5. CYBISTER

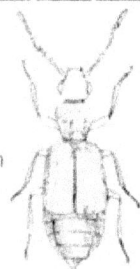

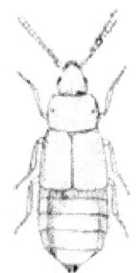

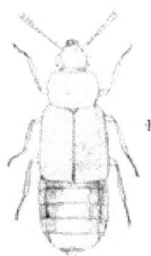

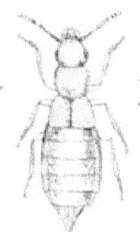

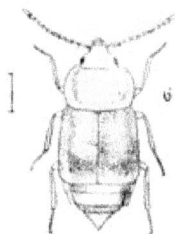

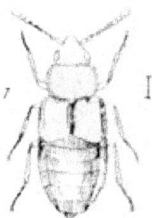

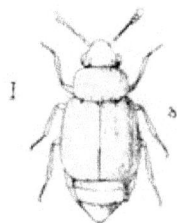

Pl. 15

1	LESTEVA.	5	MICRALYMMA
2	CORYPHIUM	6	ANTHOBIUM
3	ACIDOTA.	7	SYNTOMIUM
4	OMALIUM	8	PROTEINUS

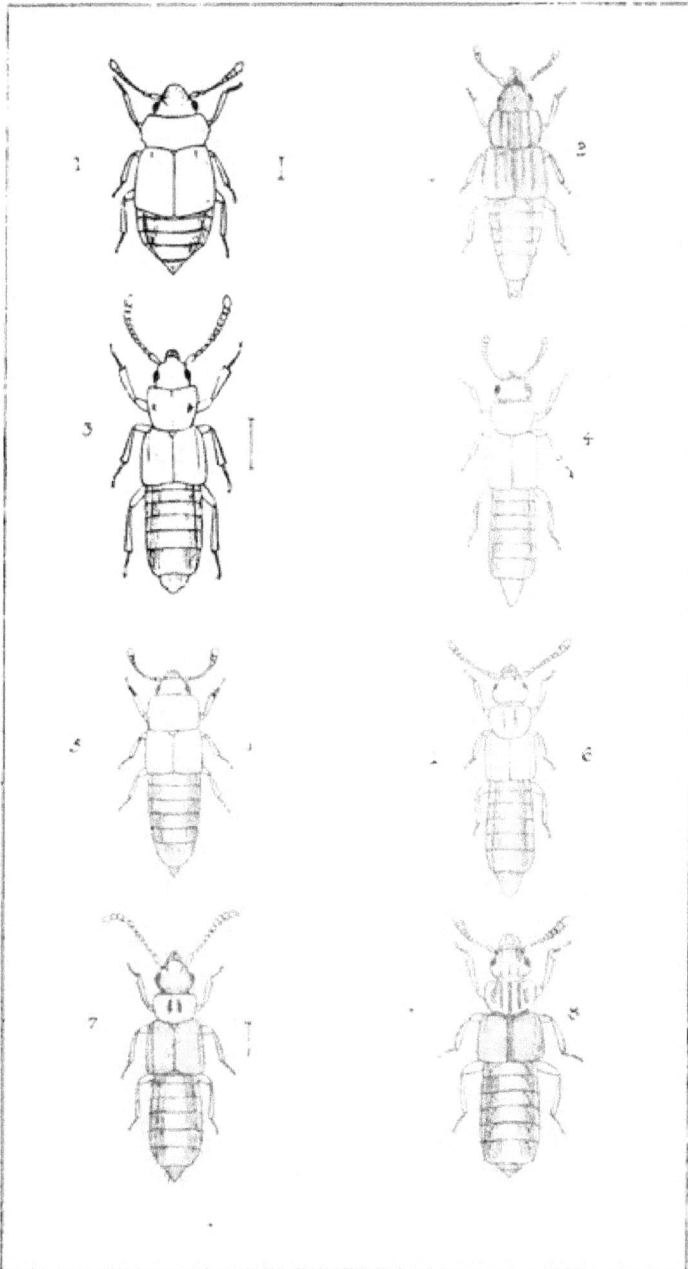

Pl 16.

1. MEGARTHRUS
2. PSEUDOPSIS.
3. COPROPHILUS.
4. TÆNOSOMA.

5. PHLÆOCHARIS
6. TROGOPHLÆUS
7. APLODERUS.
8. OXYTELUS

Pl. 17

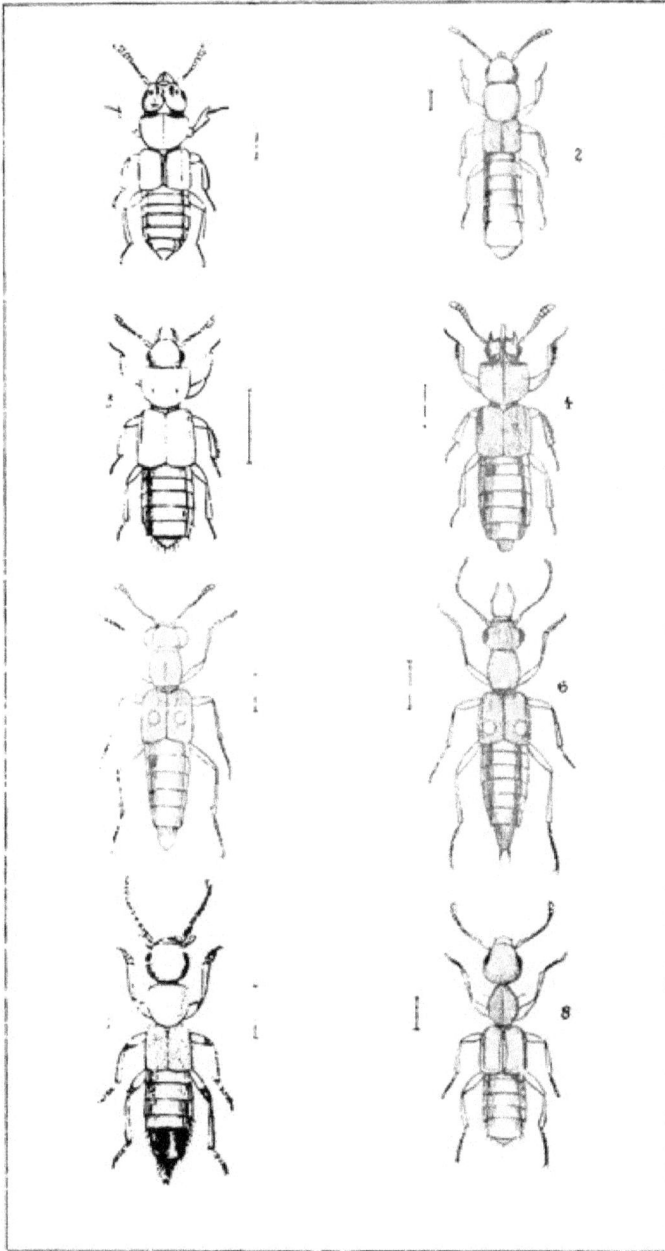

1. PLATYSTETHUS
2. PHYTOSUS
3. HESPEROPHILUS
4. BLEDIUS
5. STENUS
6. DIANOUS
7. PÆDERUS
8. RUGILUS

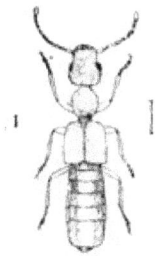

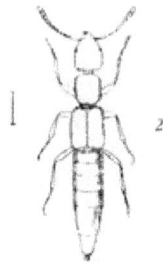

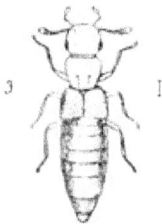

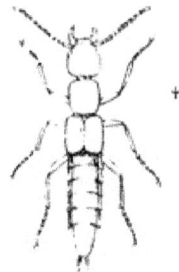

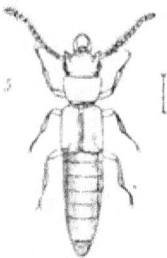

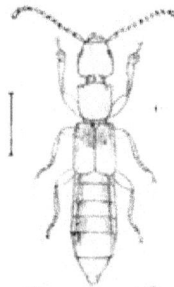

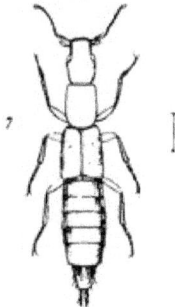

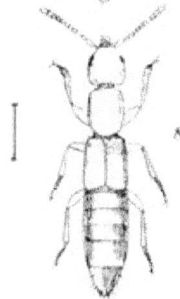

1. ASTENUS.
2. SUNIUS.
3. EVÆSTHETUS.
4. MEDON.

5. SIACONIUM.
6. ACHENIUM.
7. CRYPTOBIUM.
8. LATHROBIUM.

Pl. 18

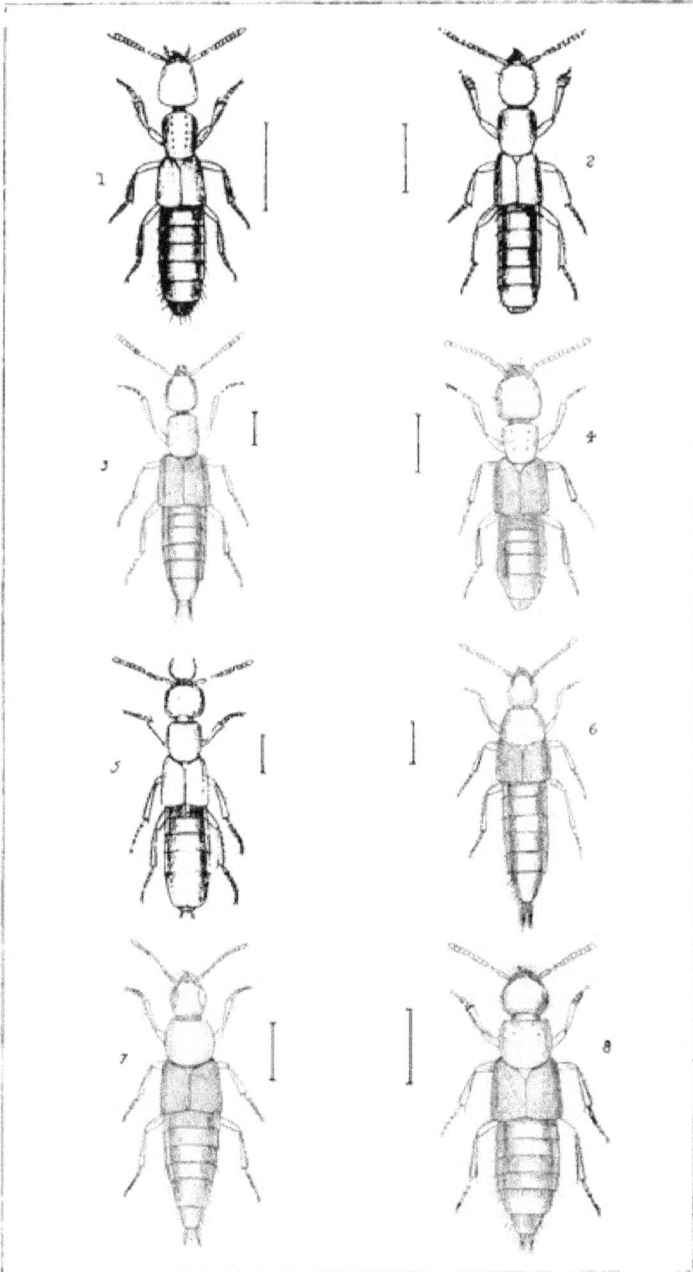

1. GYROHYPNUS.
2. OTHIUS
3. GABRIUS.
4. CAFIUS.
5. BISNIUS
6. HETEROTHOPS
7. RAPHIRUS
8. PHILONTHUS

Pl 20

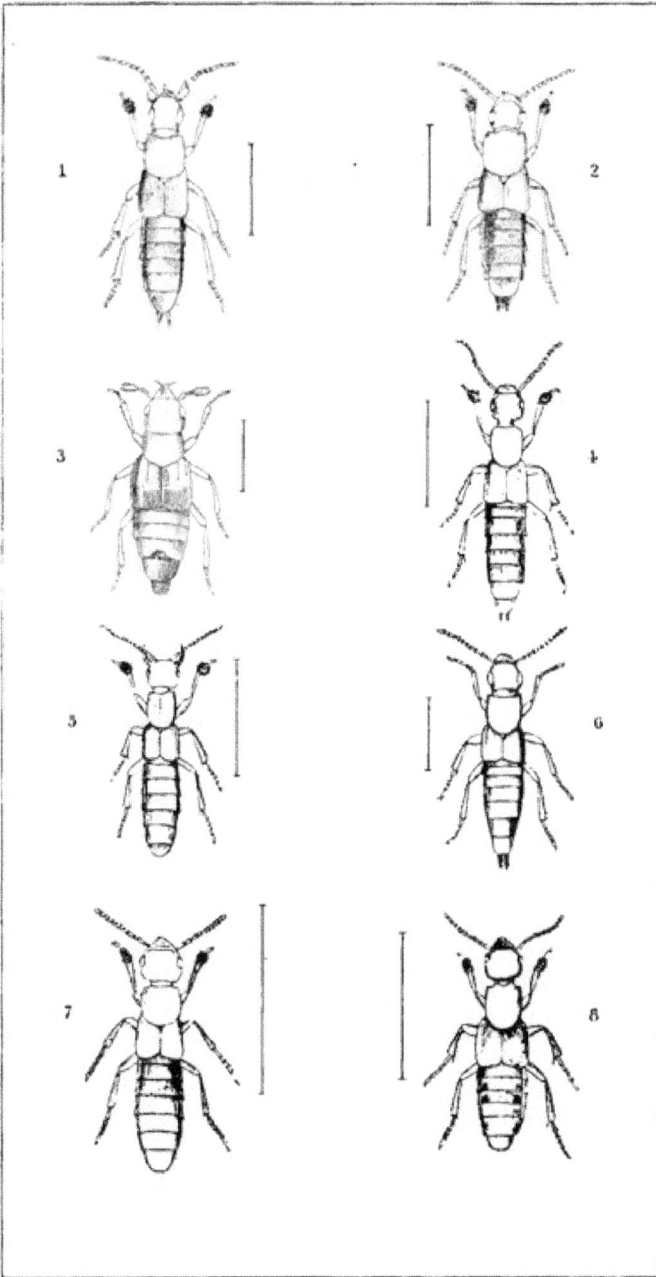

1 QUEDIUS   5 OCYPUS
2 ASTRAPÆUS   6 EURYPORUS
3 OXYPORUS   7 GOERIU⁻
4 TASGIUS   8 STAPHYLINUS

Pl 21

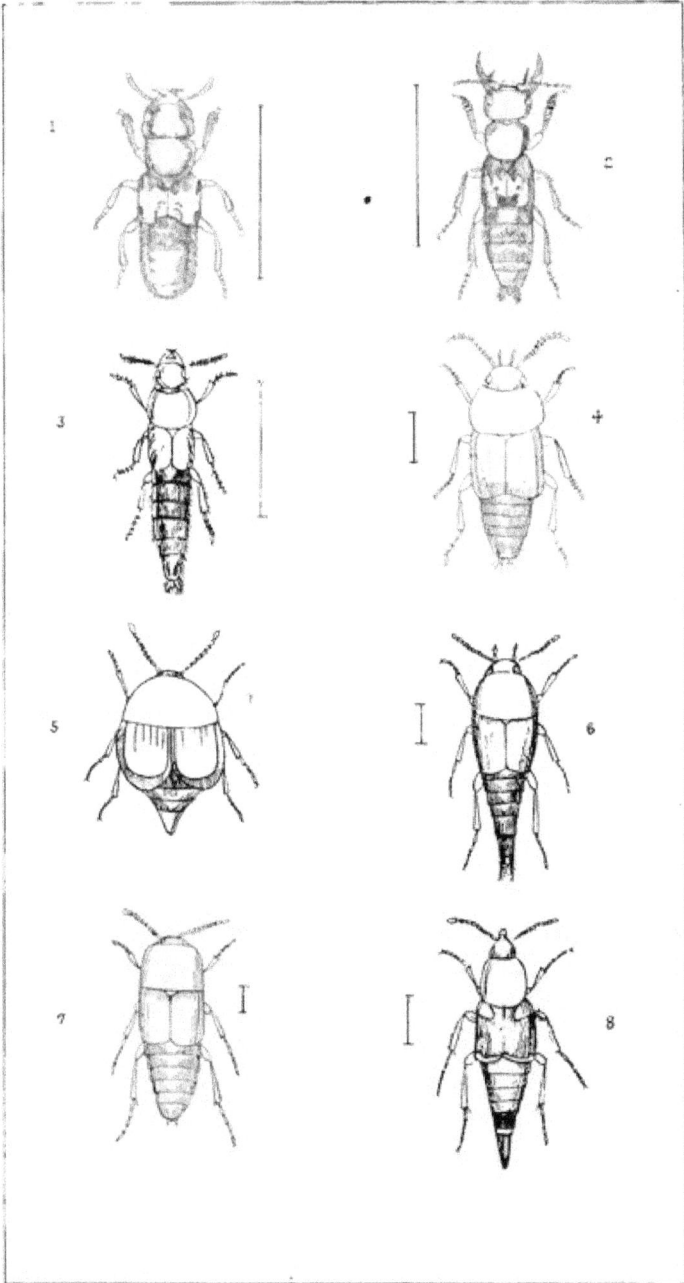

1 EMUS
2 CREOPHILUS
3 VELLEUS
4 TACHINUS

5 CYPHA
6 CONURUS
7 TACHYPORUS
BOLITOBIUS

Pl 22

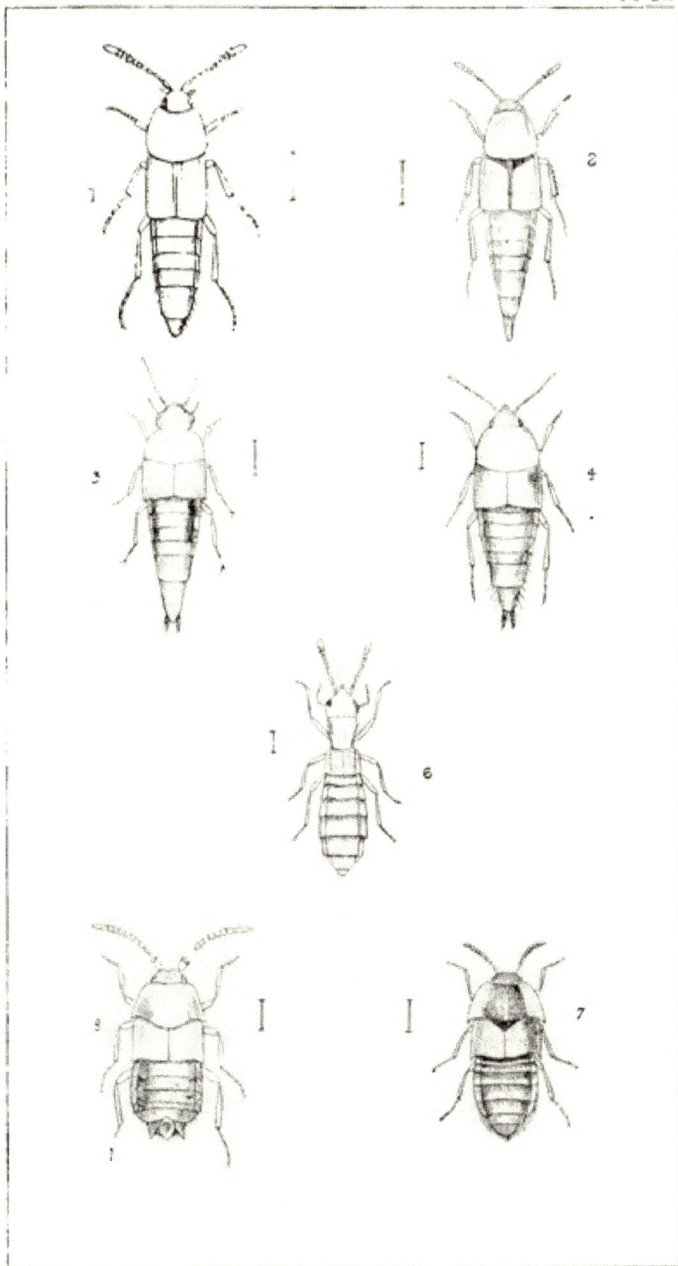

1. MEGACRONUS 4. CENTROGLOSSA.
2 MYCETOPORUS 6 DIGLOSSA.
3 DEINOPSIS 7. DINARDA
8. ATEMELES.

Pl 83.

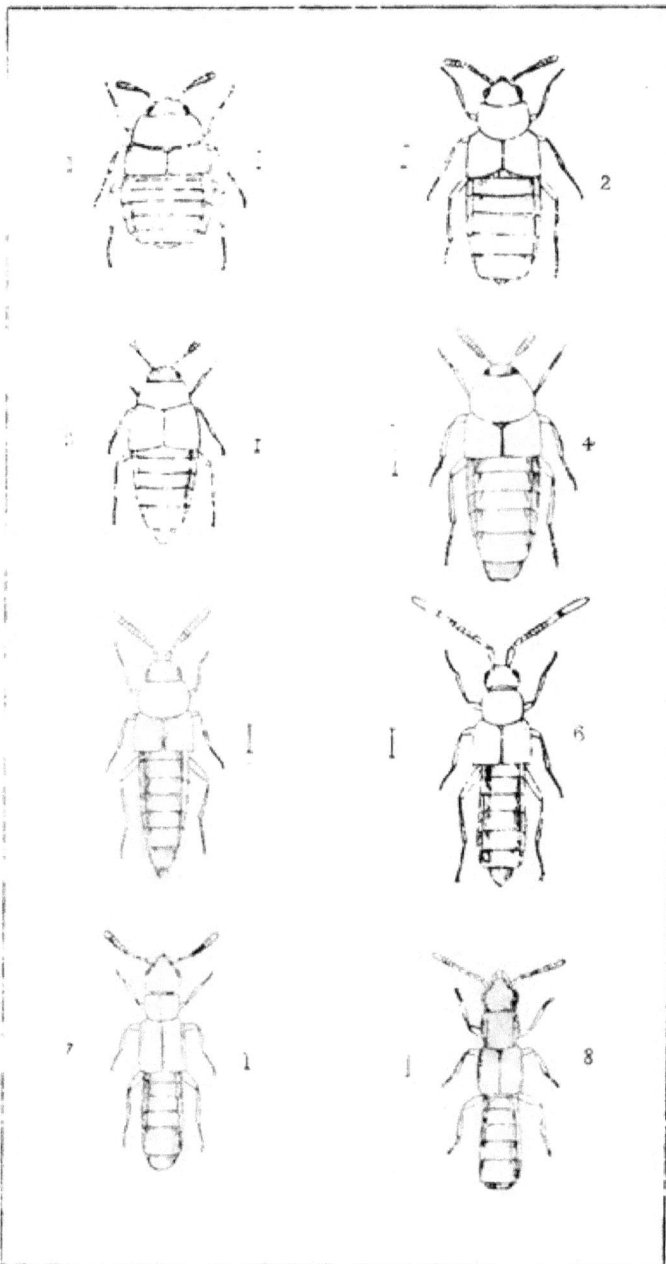

1. ENCEPHALUS.
2. GYROPHÆNA
3. OLIGOTA
4. ALEOCHARA.

5. OXYPODA.
6. CALLICERUS.
7. HOMALOTA
8. HYGRONOMA.

Pl. 24

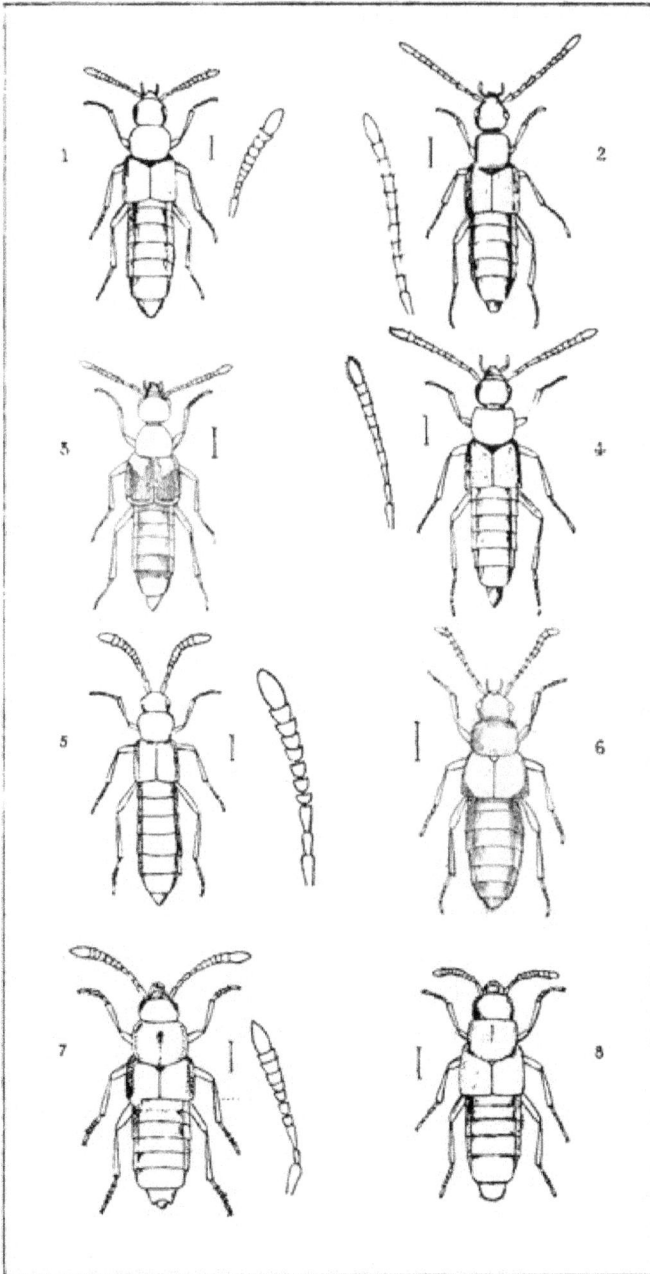

1 PHLŒOPORA.
2 TACHYUSA
3 BOLITOCHARA.

5 CALODERA
3 ZYRAS.
7 PELLA.

Pl 25.

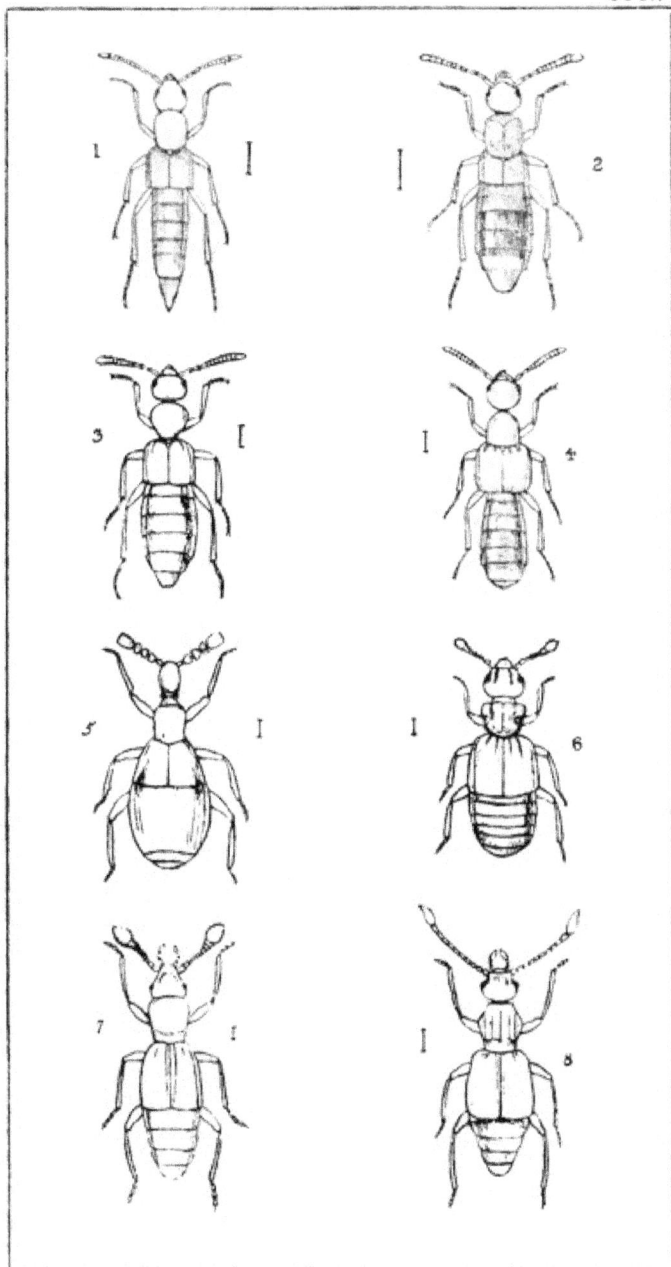

1 ISCHNOPODA
2 ASTILBUS
3 FALAGRIA
4 AUTALIA

5 CLAVIGER
6 EUPLECTUS
7 TRIMIUM
8 BATRISUS

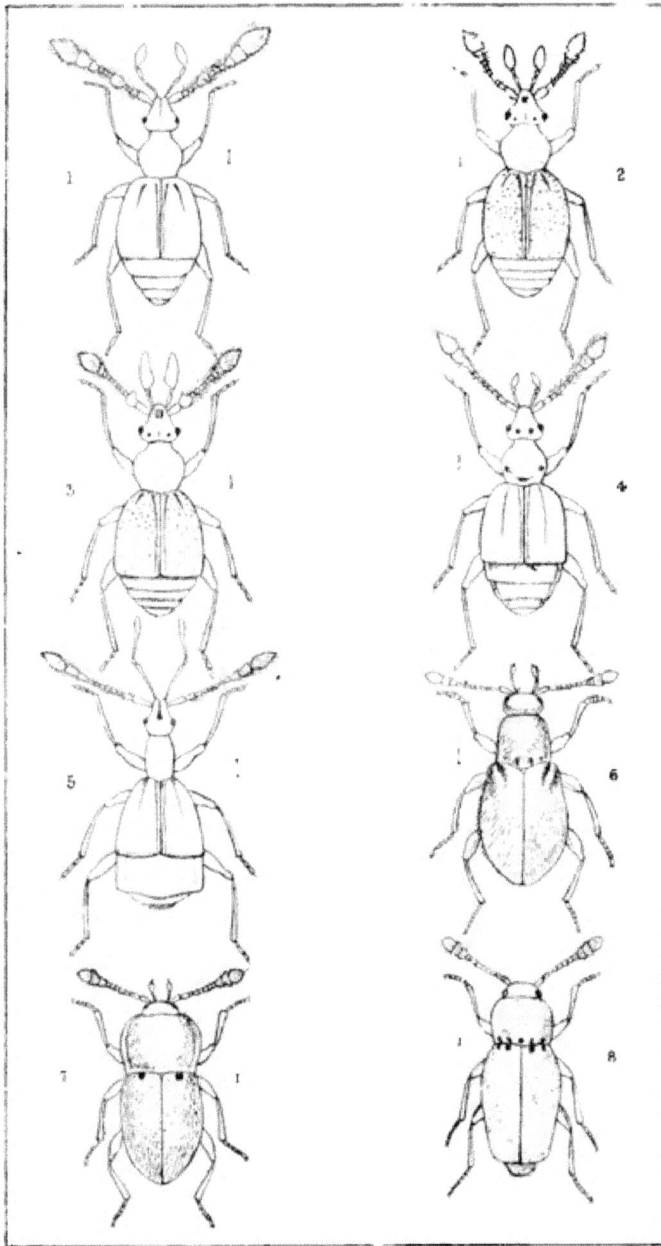

Pl.26

1. TYCHUS
2. ARCOPAGUS.
3. BYTHINUS.
4 BRYAXIS.

5. PSELAPHUS.
6. SCYDMÆNUS.
7. MEGALADERUS.
8. EUTHEIA.

Pl 27

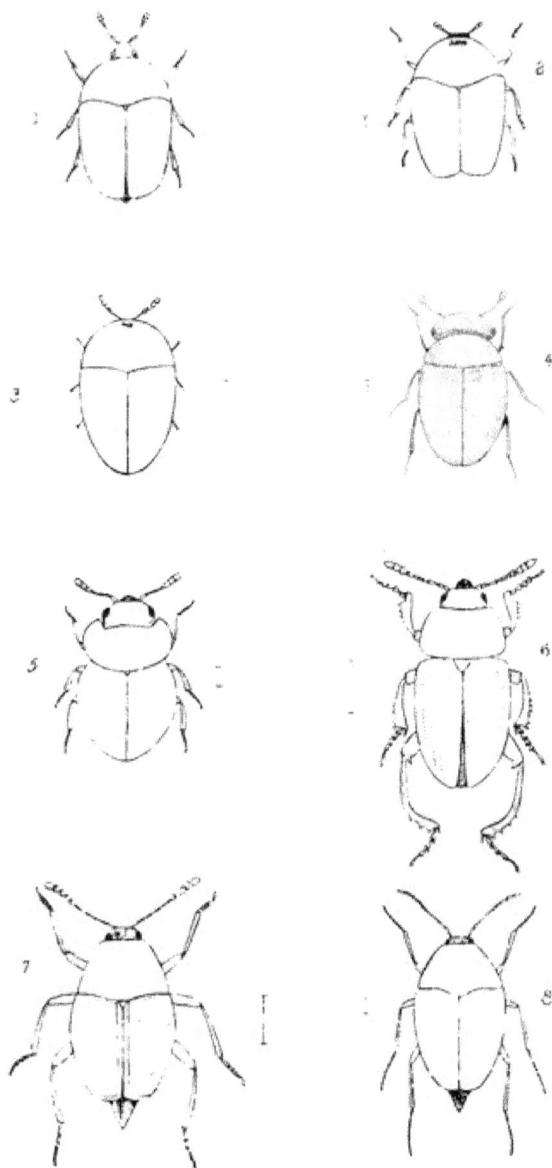

1 SERICODERUS
2 ORTHOPERUS
3 CLYPEASTER
4 CLAMBUS

5 ACATHIDIUM
6 LEIODES
7 SCAPHIDIUM
8 SCAPHISOMA

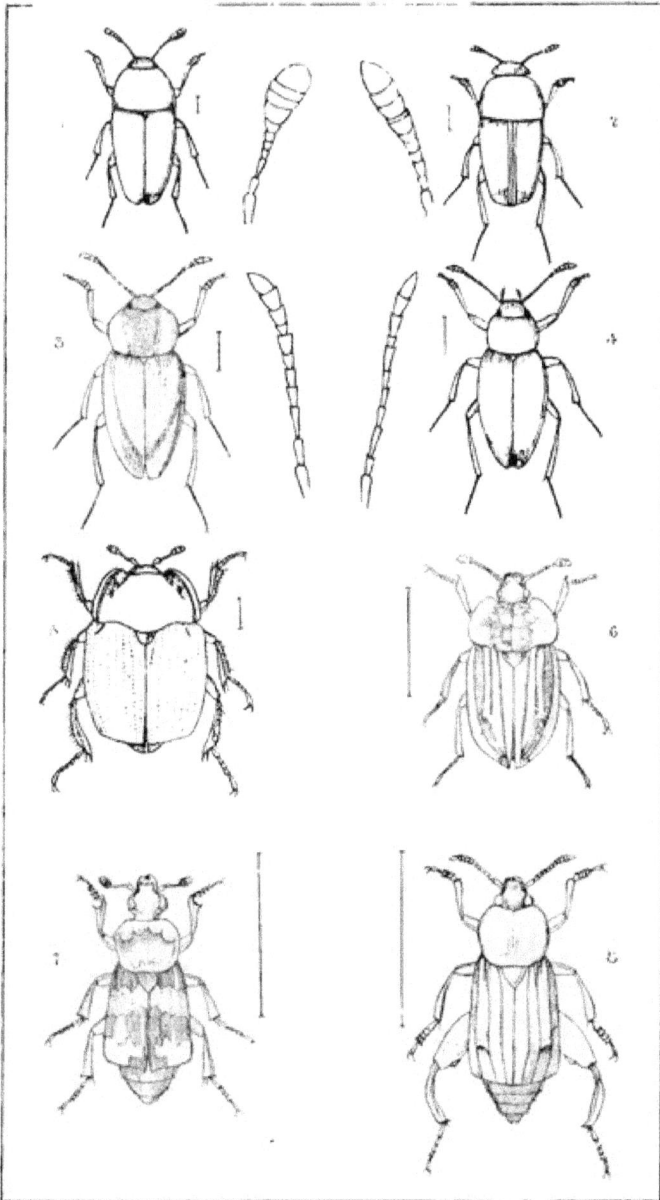

1. MYLÆCHUS (COLON)
2. PTOMAPHAGUS.
3 CATOPS.
4 CHOLEVA

5 SPHÆRITES.
6 NECROPHORUS.
7 NECRODES.
8 OICEOPTOMA.

Pl 29

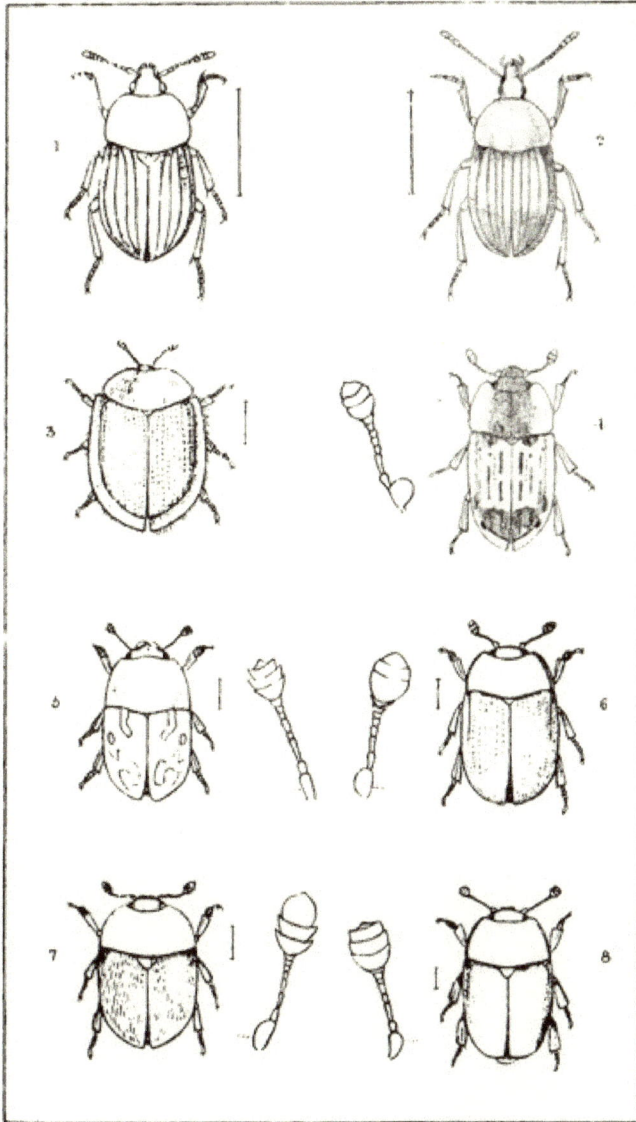

1 SILPHA.
2 PHOSPHUGA.
3 THYMALUS.
4 NITIDULA.
5. CRYPTARCHUS.
6. STRONGYLUS.
7. CAMPTA.
8. MELIGETHES.

Pl 30.

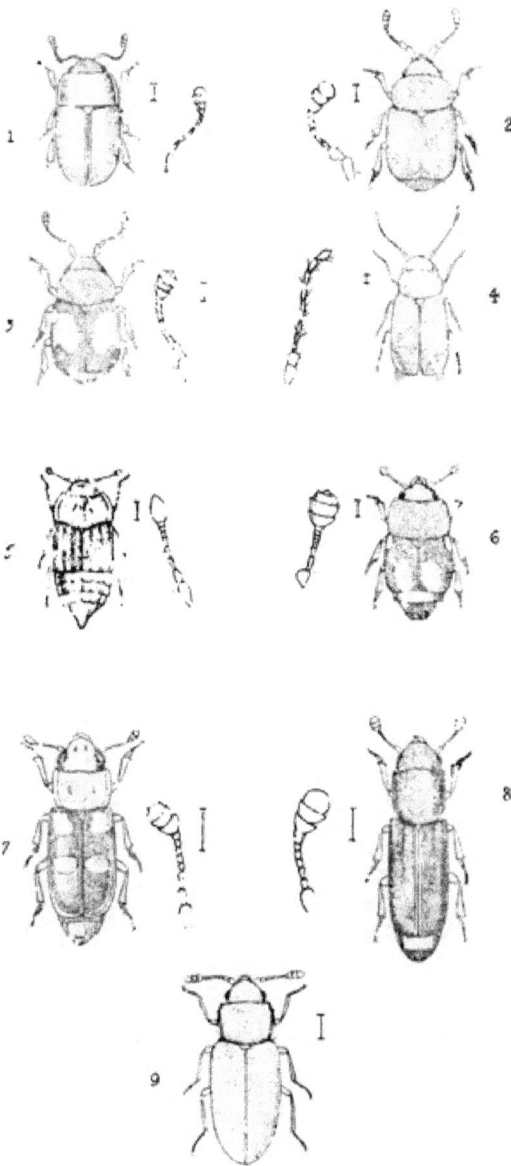

1 PRIA
2 ANOMŒOCERA
3 CATERETES
4 TRICHOPTERYX.
5 MICROPEPLUS
6 CARPOPHILUS
7 IPS
8 PITYOPHAGUS
9 CRYPTOPHAGUS.

Pl 81

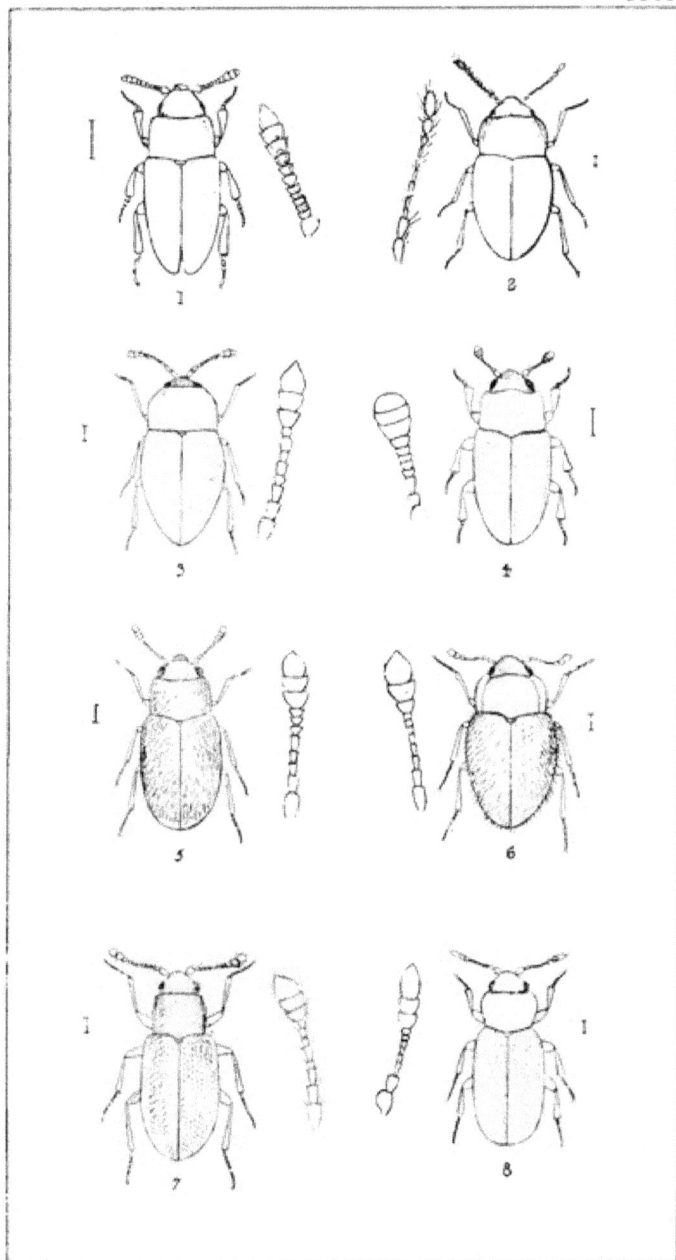

1. ANTHEROPHAGUS
2. ANISARTHRIA.
3. ATOMARIA
4. ENGIS.

5. TYPHÆA
6. MYCETÆA
7. PARAMECOSOMA
8. CORTICARIA.

Pl 52

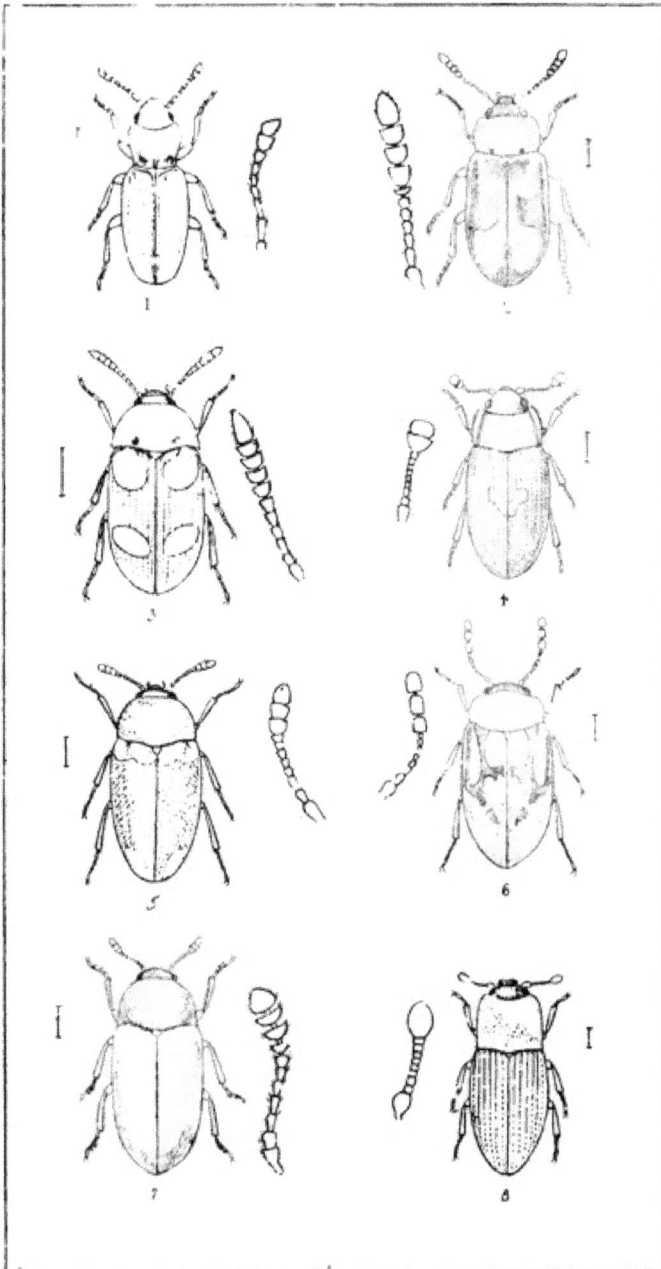

1. HOLOPARAMECUS
2. FETRATOMA.
3. MYCETOPHACUS
4. BIFHYLLUS

5. TRIPHYLLUS.
6. PHLOIOPHILUS
7. BYTURUS
8. CERYLON

Pl 33

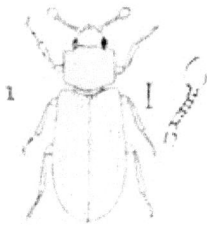

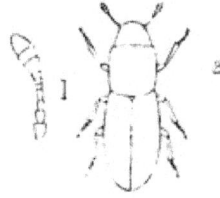

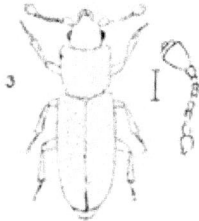

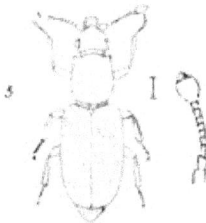

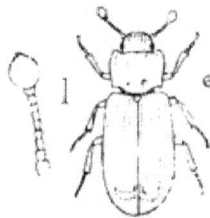

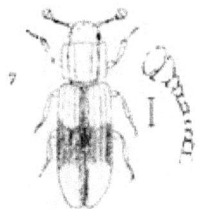

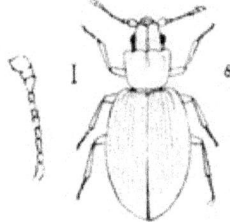

1 SYNCHITA.
2 ANOMMATUS.
3 RHYZOPHAGUS.
4 LISSODEMA

5 MONOTOMA.
6 CICONES.
7 BITOMA.
8 LATRIDIUS.

Pl 34

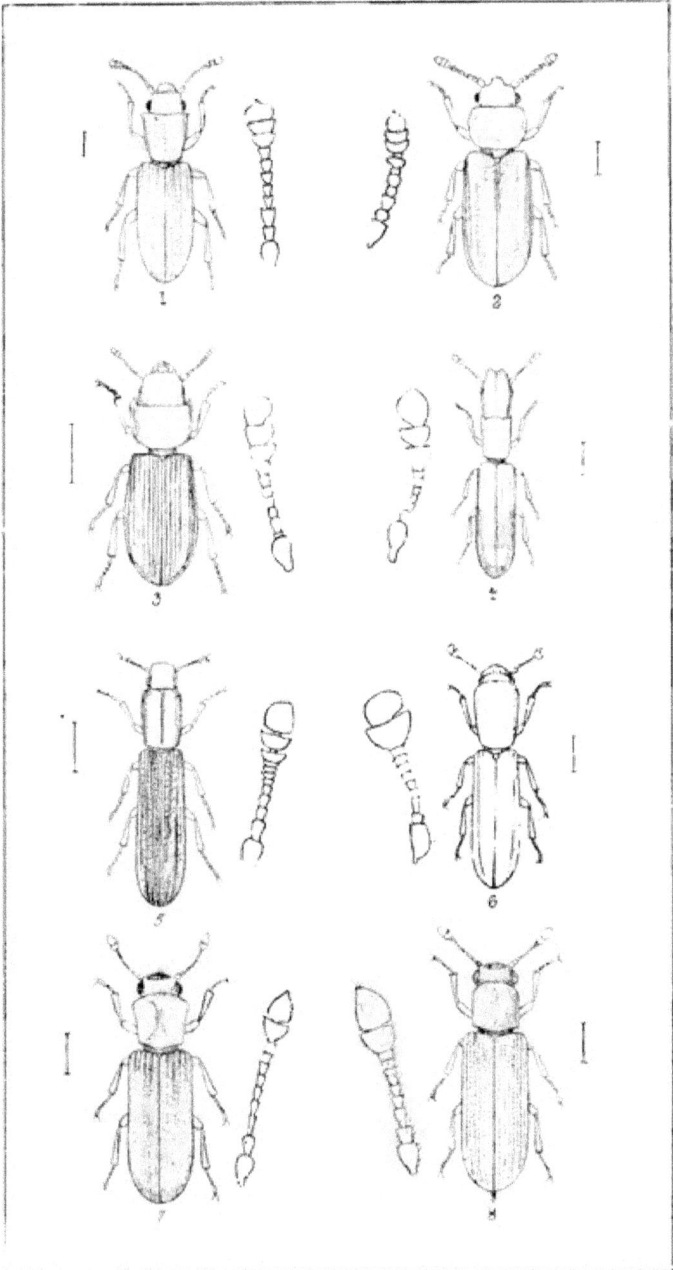

1. SILVANUS   5. COLYDIUM
2. PEDIACUS   6. TEREDUS

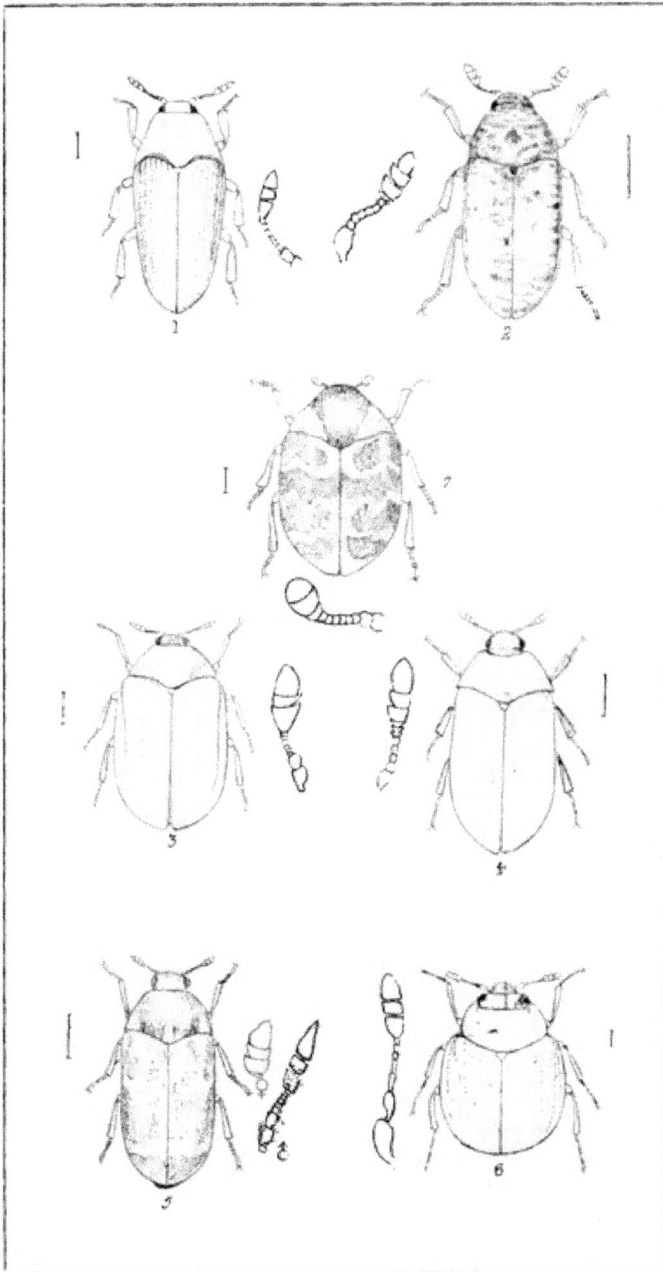

1. THROSCUS 4 ATTAGENUS
2. DERMESTES 5 MEGATOMA
3. TIRESIAS 6 ASPIDIPHORUS
. 7 ANTHRENUS

Pl. 36

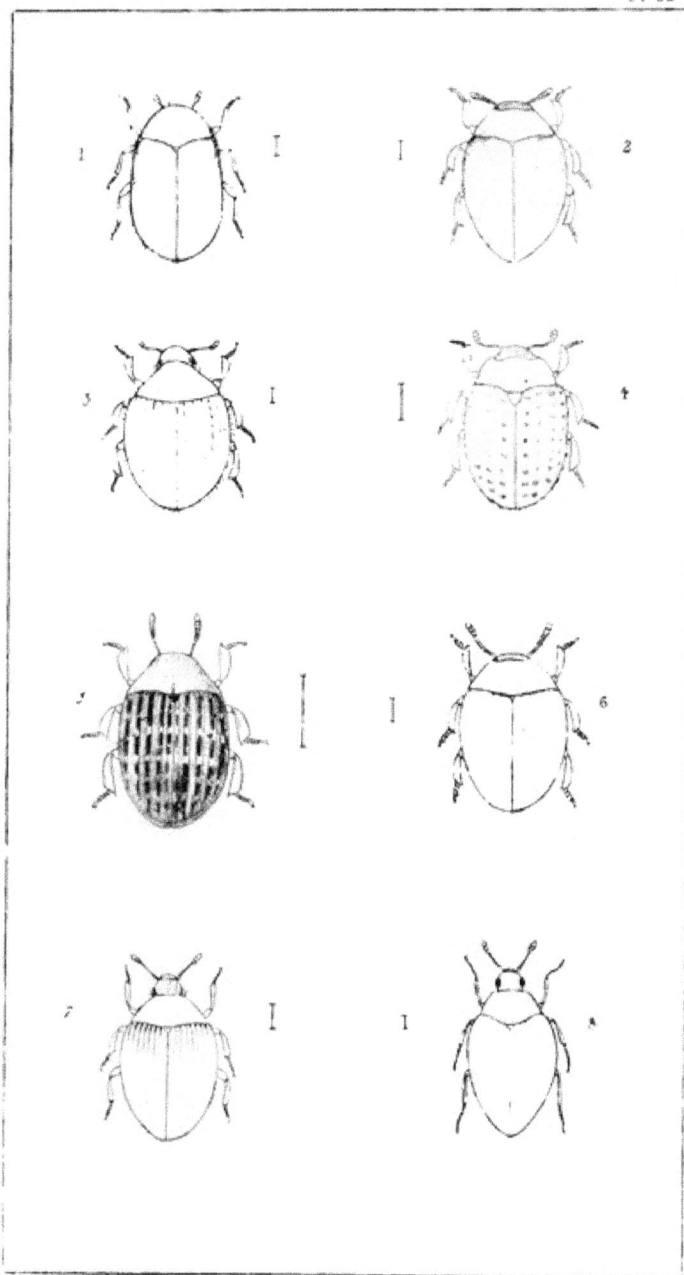

1 TRINODES
2 LIMNICHUS
3 SYNCALYPTA
4 NOSODENDRON

5 BYRRHUS
6 OOMORPHUS
7 SIMPLOCARIA
8 EPHISTEMUS

Pl 37

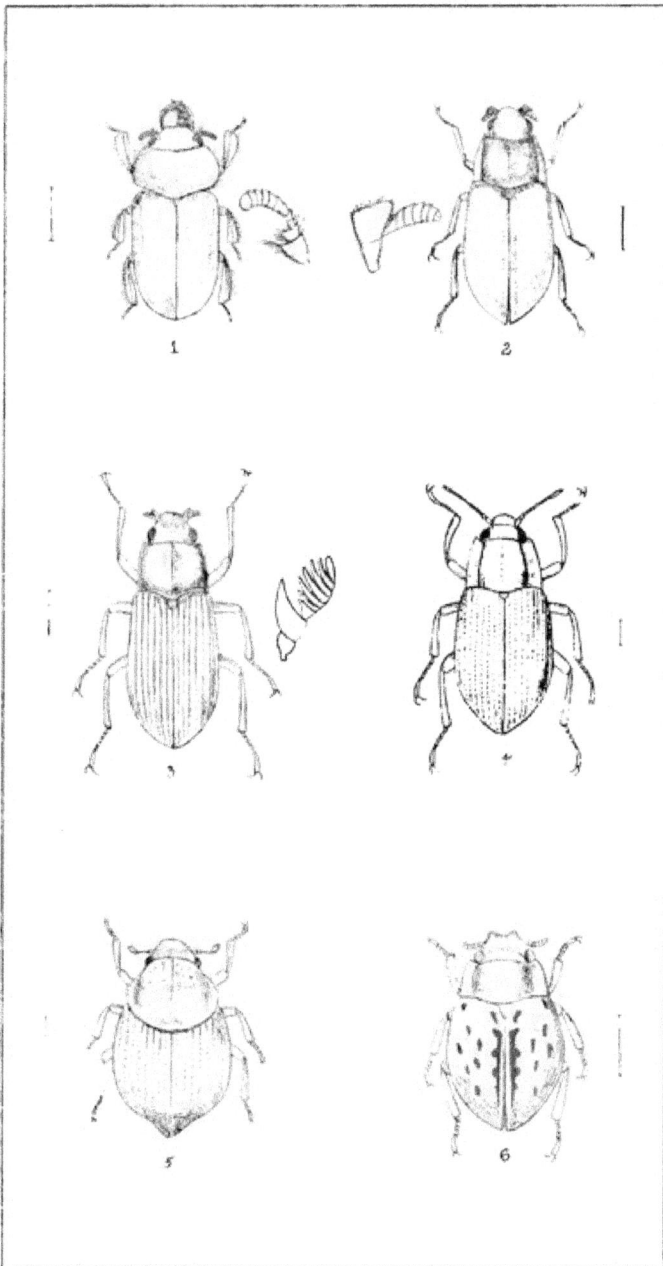

1. HETEROCERUS.
2. PARNUS.
3. DRYOPS.
4. ELMIS
5. GEORYSSUS.
6. SPERCHEUS

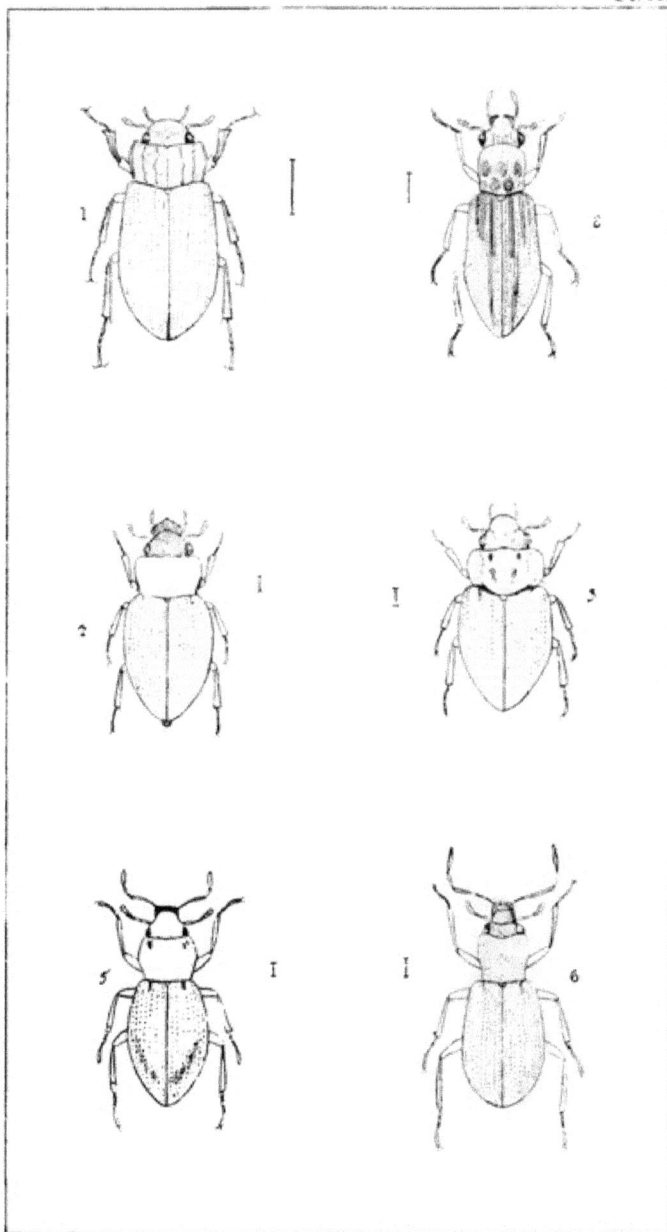

1. HELOPHORUS.
2. HYDROCHUS
3. ENICOCERUS
4. OCHTHEBIUS.
5. AMPHIBOLUS.
6. HYDRÆNA.

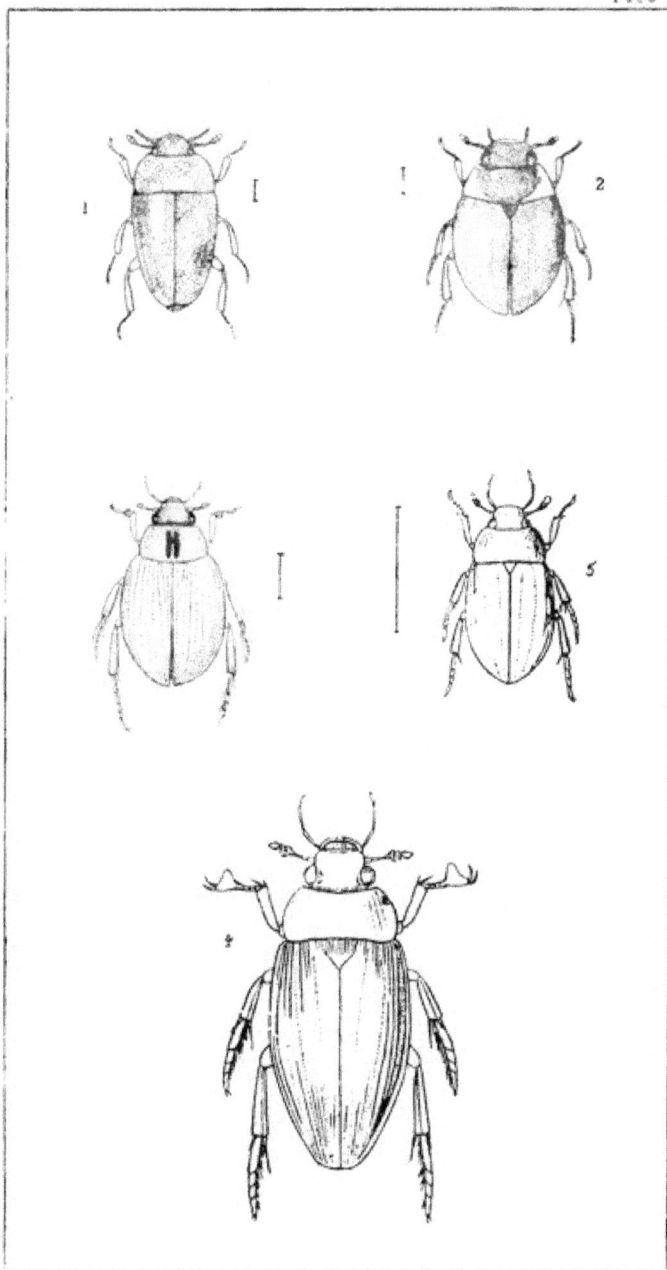

1 LIMNEBIUS 3 BEROSUS
2 LACCOBIUS 4 HYDROUS
5 HYDROPHILUS

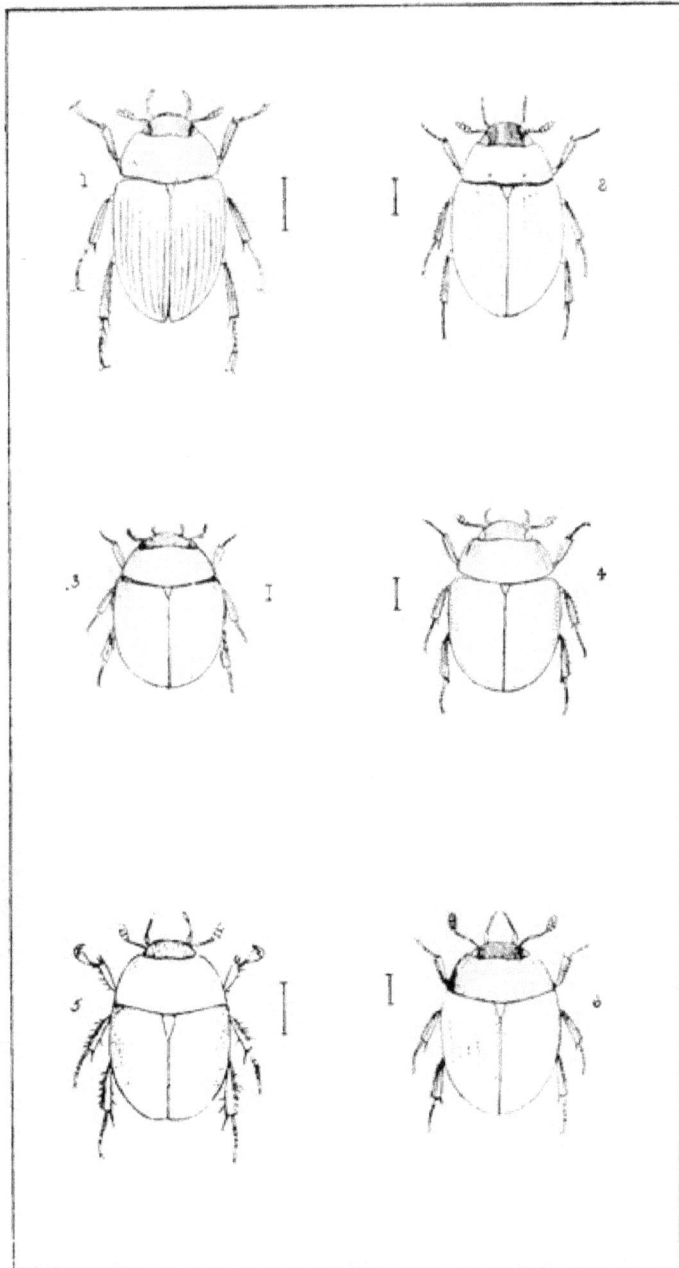

1. HYDROBIUS
2. PHILHYDRUS
3. CHÆTARTHRIA
4. CYCLONOTUM.
5. SPHÆRIDIUM
6. CERCYON.

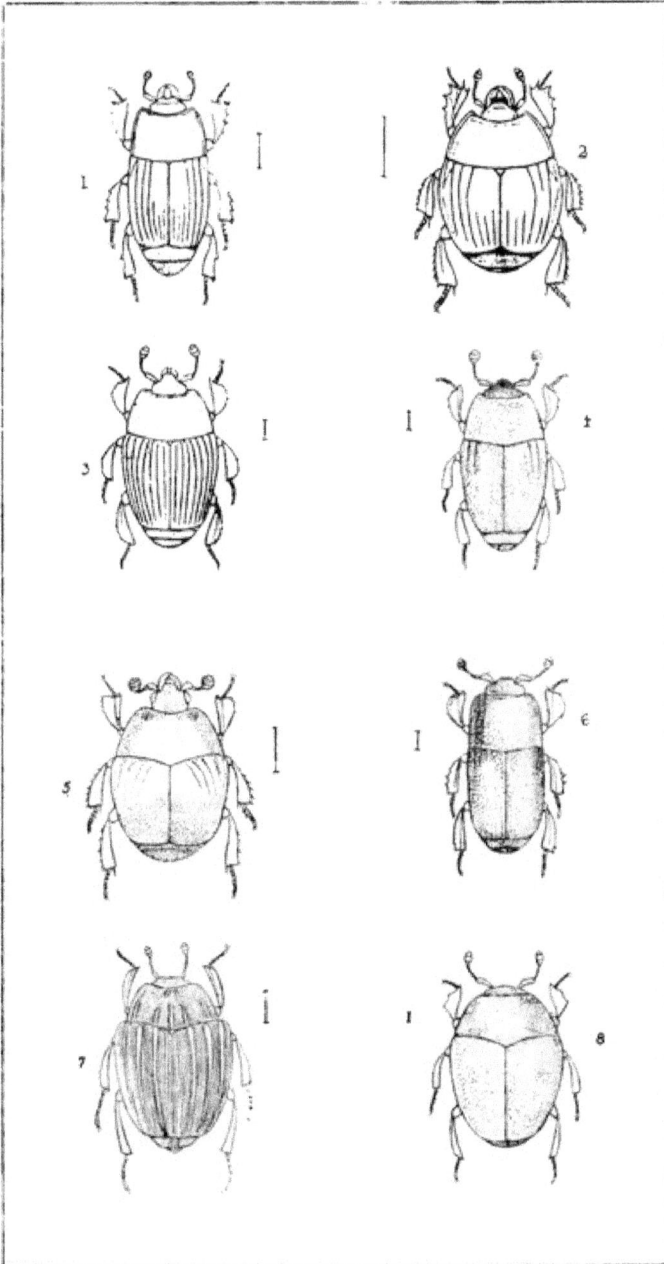

Pl 41

1. PLATYSOMA.	5 SAPRINUS.
2 HISTER.	6 TERETRIUS
3. DENDROPHILUS	7 ONTHOPHILUS.
4 PAROMALUS	8 ABRÆUS.

Pl 42.

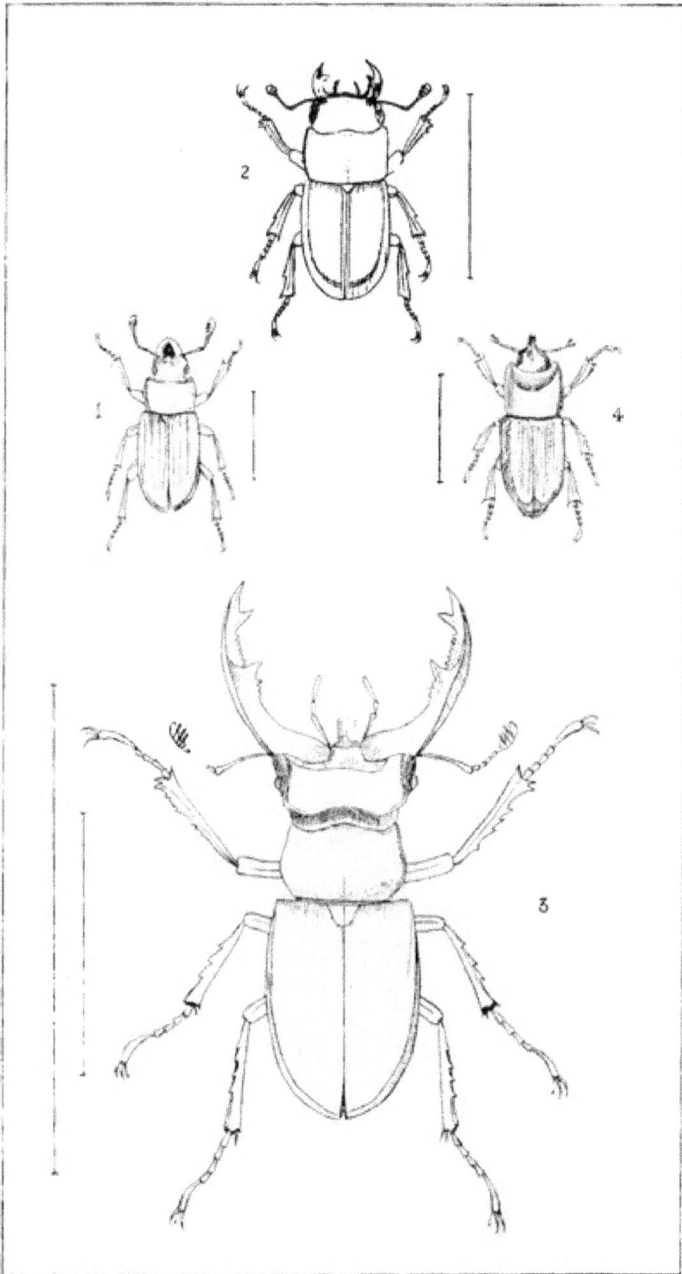

1 PLATYGERUS   3 LUCANUS

2 DORCUS   4 SINODENDRON

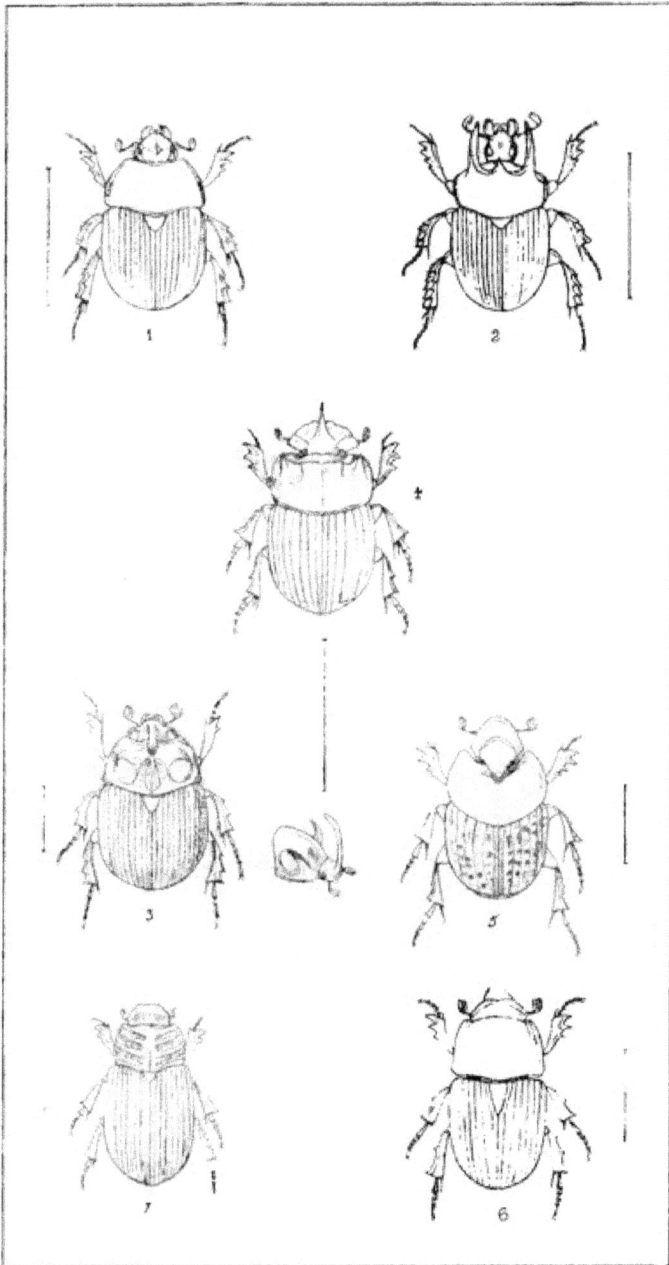

1 GEOTRUPES　　　4 COPRIS
2 TYPHŒUS　　　5 ONTHOPHAGUS
3 BOLBOCERUS　　6 APHODIUS
" PSAMMODIUS

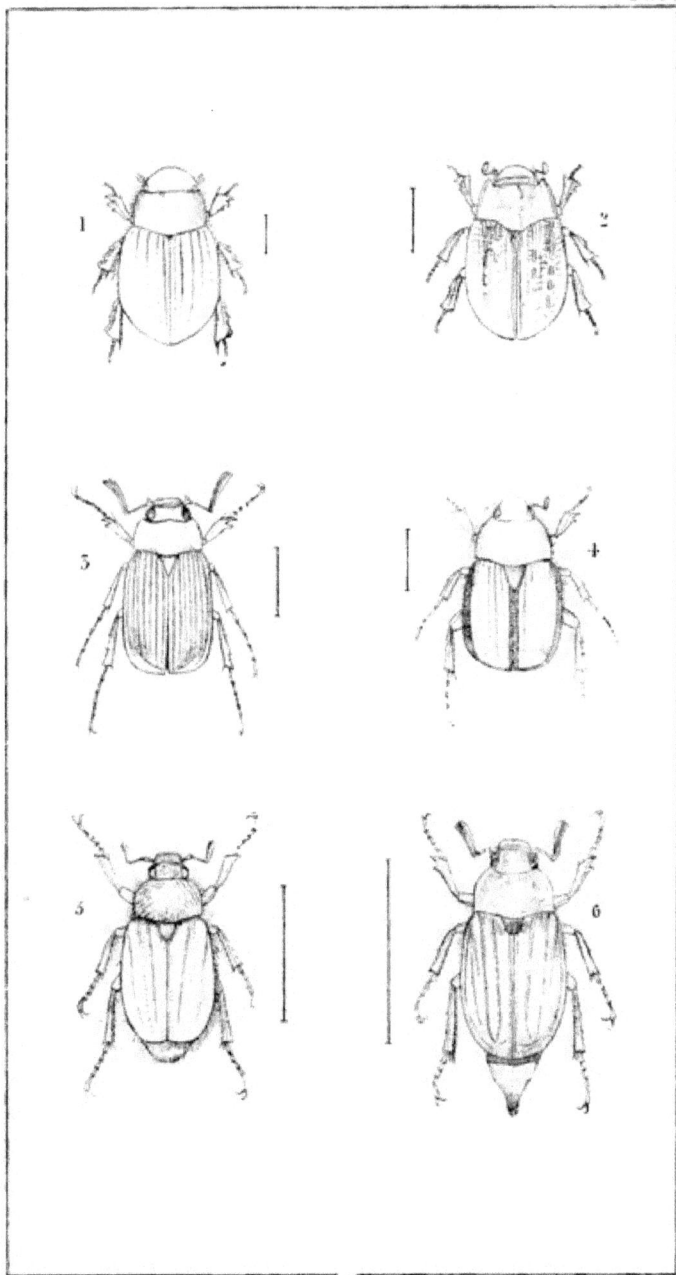

Pl.4

1. ÆCIALIA .
2. TROX .
3. SERICA .

4. OMALOPLIA .
5. RHISOTROCUS .
6. MELOLONTHA .

Pl 45

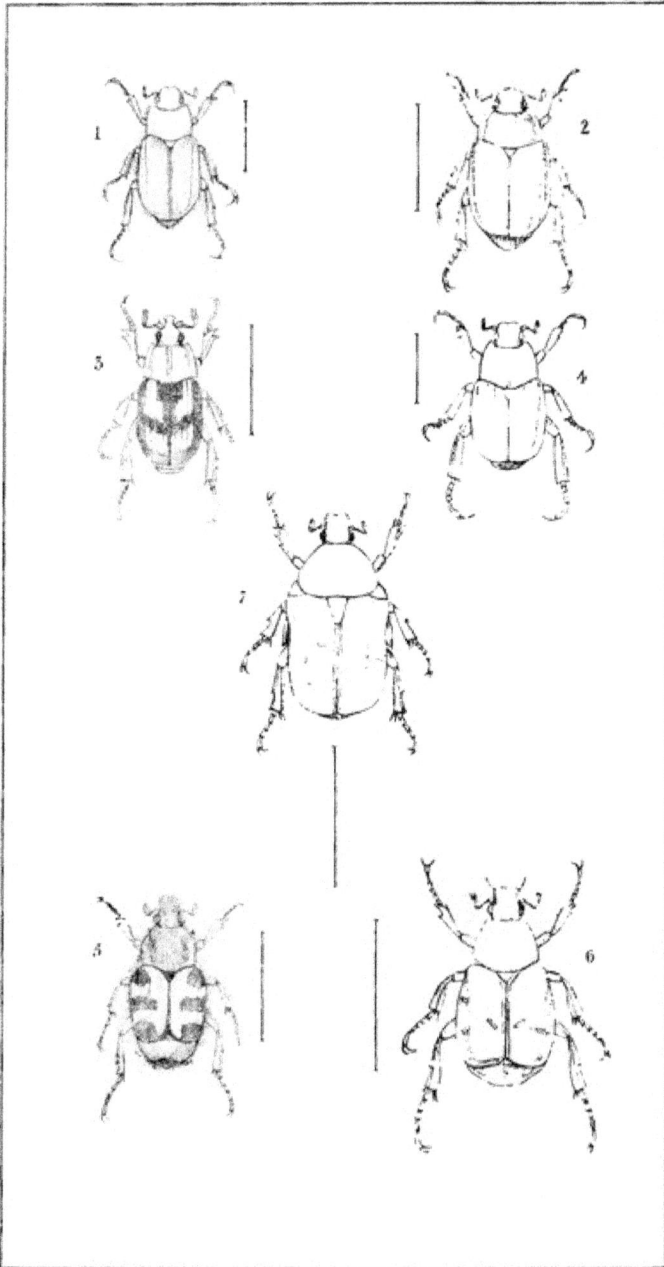

1. PHYLLOPERTHA.   4 HOPLIA.
2. ANOMALA.   5. TRICHIUS.
3. ANISOPLIA.   6. GNORIMUS.
7. GETONIA.

Pl 46

1 ANTHAXIA
2 AGRILUS
3 APHANISTICUS
4 TRACHYS

5 MELASIS
6 MICRORHAGUS
7 ADRASTUS
8 AGRIOTES

Pl 42.

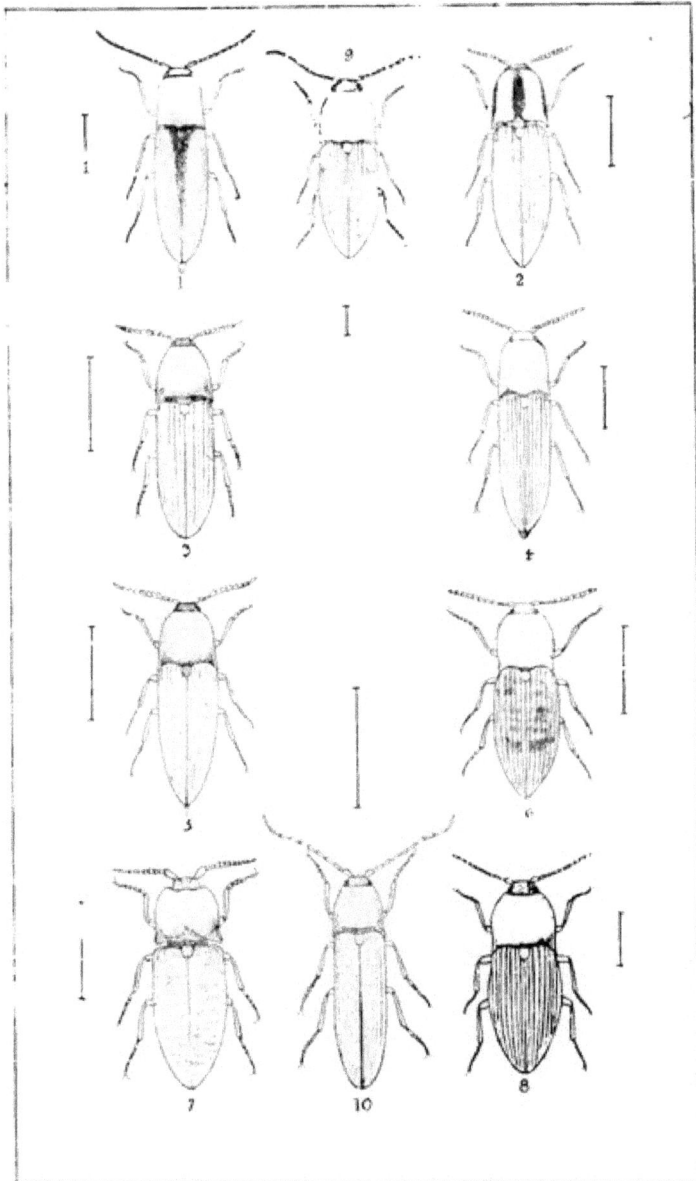

1 DOLOPIUS ·
2 SERICOSOMUS.
3 ECTINUS.
4 LIMONIUS
5 ELATER

6 PROSTERNON
7 AGRYPNUS
8 HYPOLITHUS.
9 CRYPTOHYPNUS
10. MELANOTUS

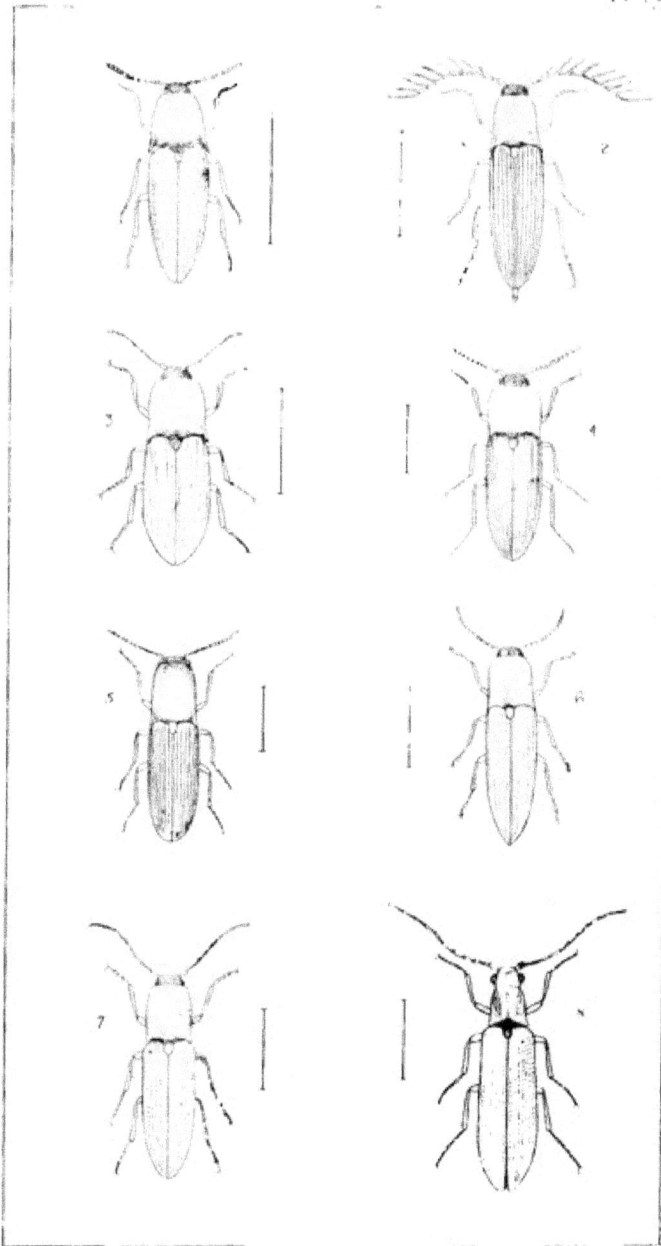

1 LUDIUS	5 APLOTARSUS
2 CTENICERUS	6 CTENONYCHUS
3 SELATOSOMUS	7 ATHOUS
4 CARDIOPHORUS	8. CAMPYLUS

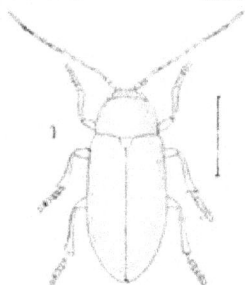

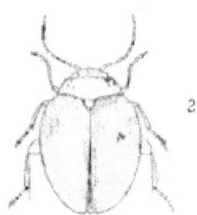

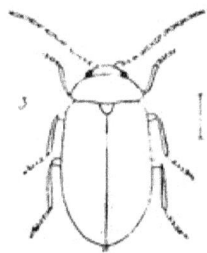

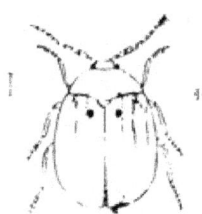

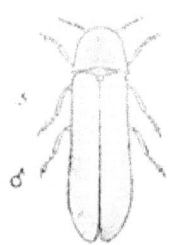

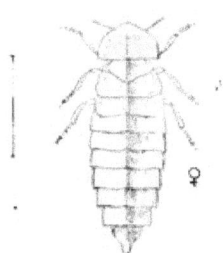

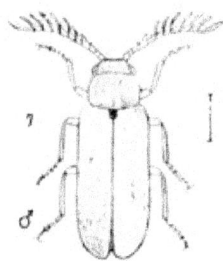

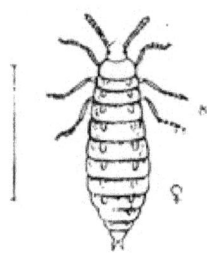

1 ATOPA   5. LAMPYRIS ♂
2 SCIRTES   6 FEMALE ♀
3 CYPHON   7 DRILUS ♂
4 EUBRIA   8 FEMALE ♀

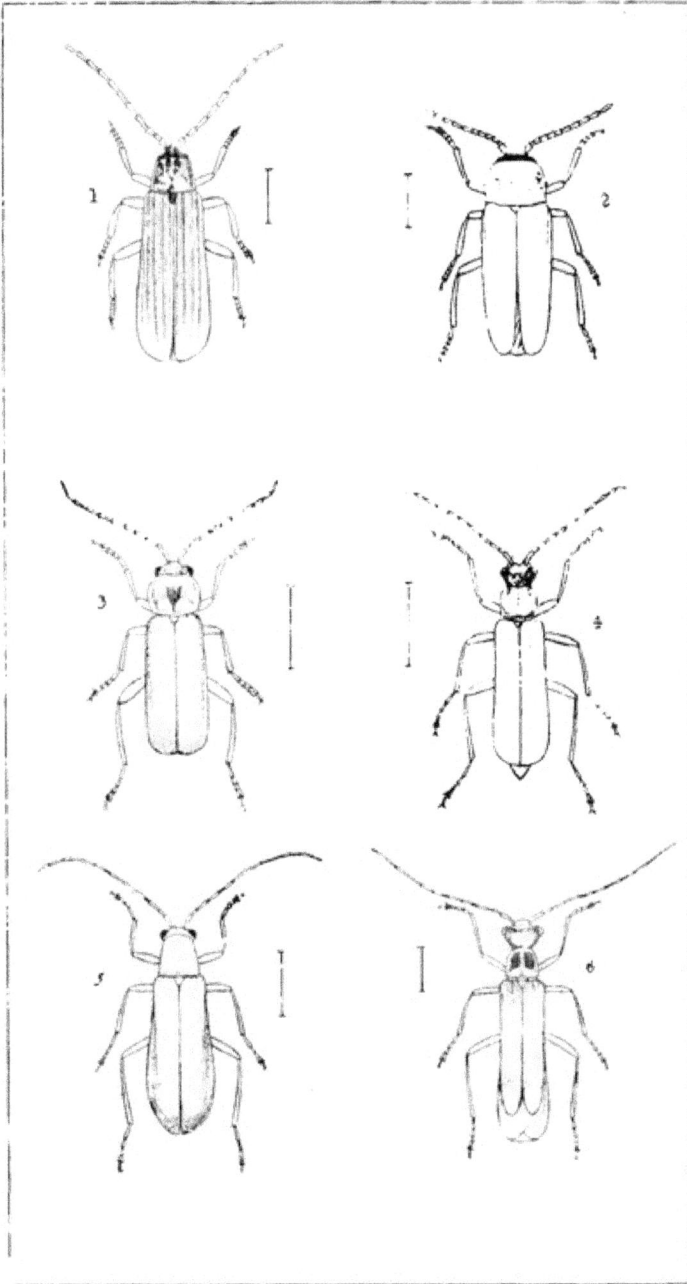

Pl 50

1 DICTYOPTERA.
2 SILIS.
3 TELEPHORUS

4 PODABRUS
5 RAGONYCHA
6 MALTHINUS

Pl. 51

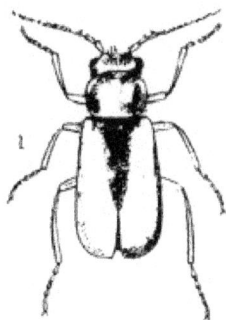

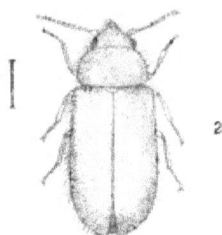

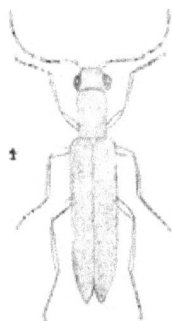

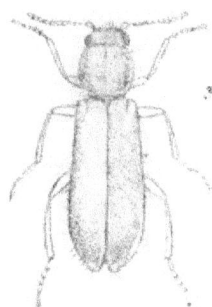

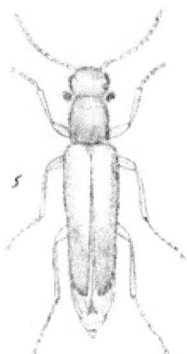

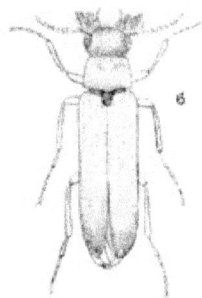

1. MALACHIUS.
2. APLOCNEMUS.
3. DASYTES.

4. DOLICHOSOMA
5. LYMEXYLON
6. HYLECÆTUS

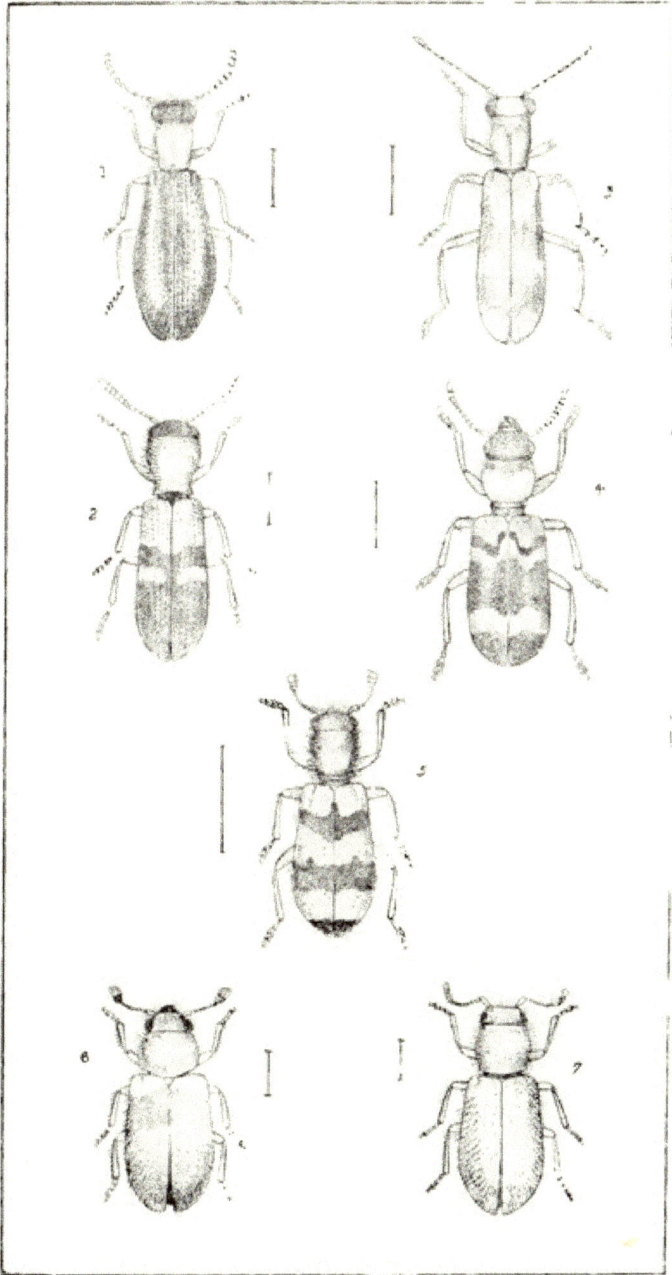

1 TILLUS.   4 THANASIMUS

2. TILLOIDEA.   5 CLERUS

3. OPILUS.   6 NECROBIA

7 CORYNETES

Pl 53

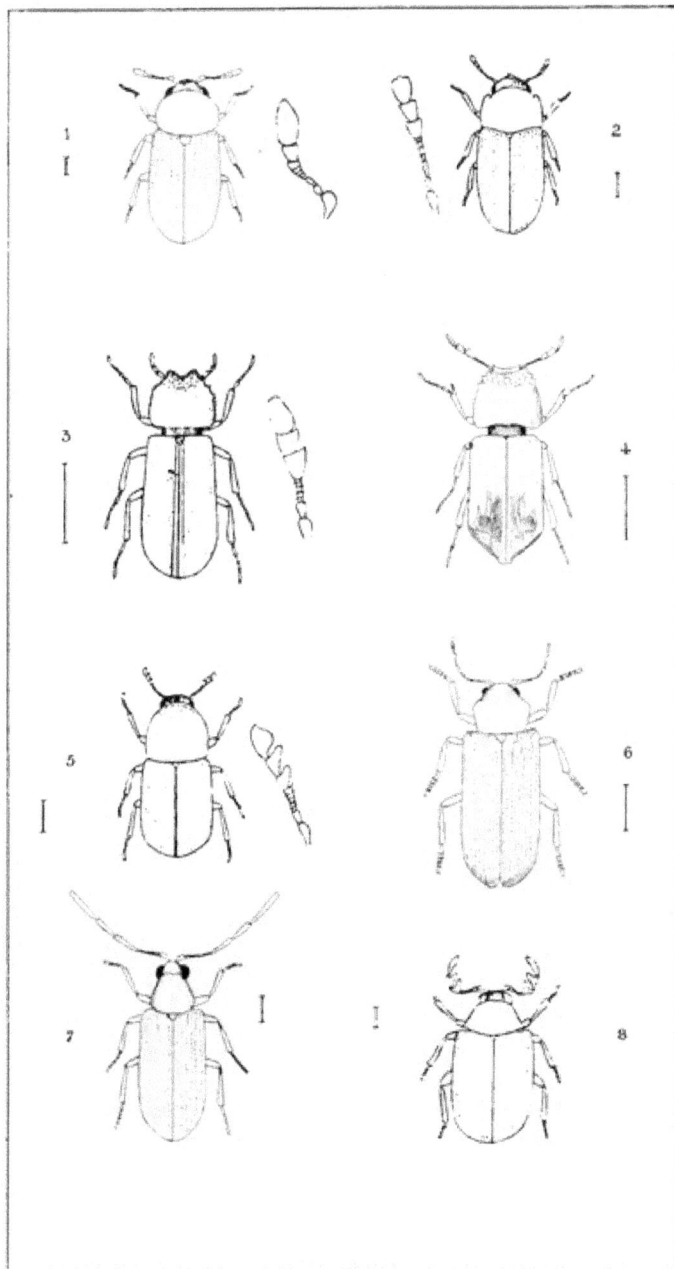

1. SPHINDUS
2. CIS
3. BOSTRICHUS.
4. APATE.
5. DINODERUS
6. ANOBIUM.
7. DRYOPHILUS
8. DORCATOMA

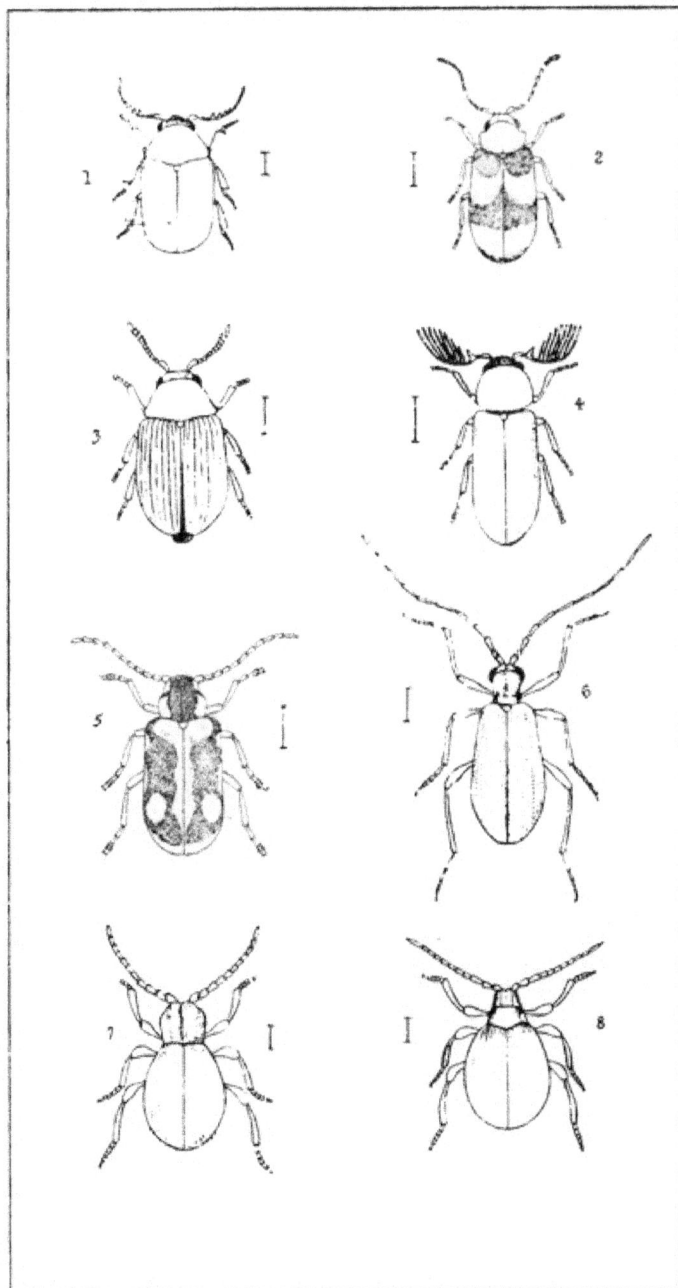

Pl. 54.

1 LASIODERMA.   5 HEDOBIA.
2 OCHINA.   6 PTINUS.

Pl. 55

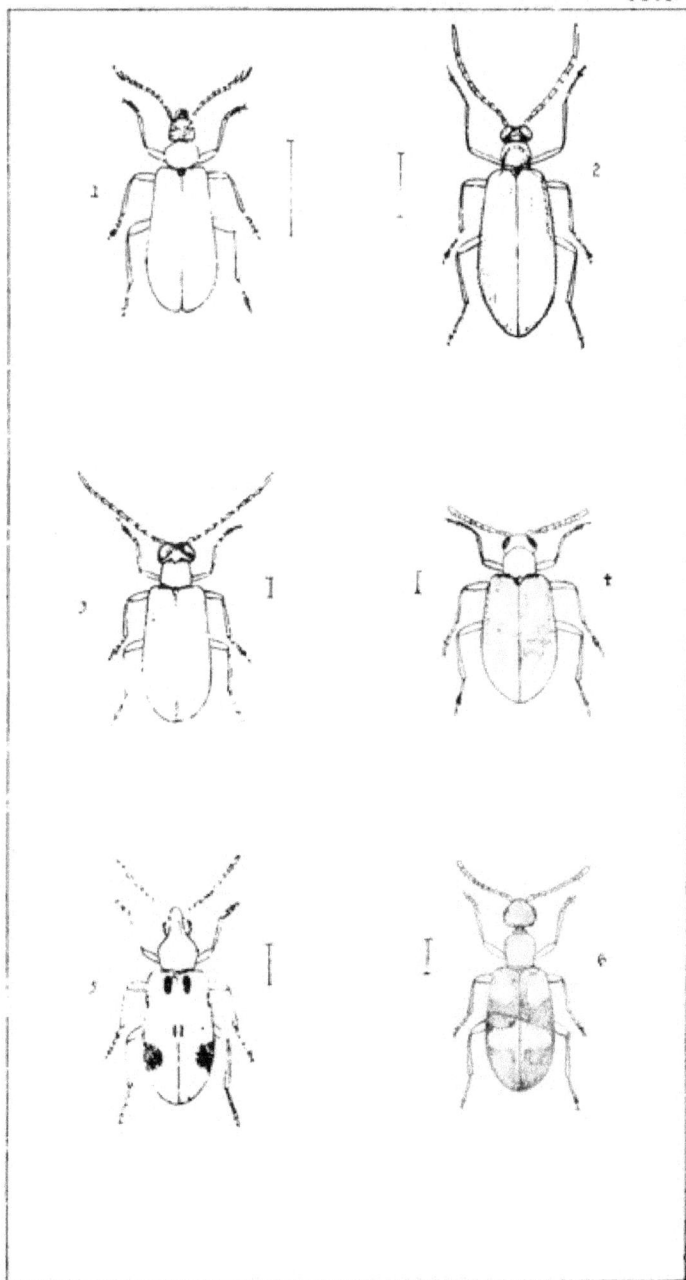

1 PYROCHROA   4 ADERUS
2 LAGRIA   5 NOTOXUS
3 XYLOPHILUS   6. ANTHICUS

Pl. 56

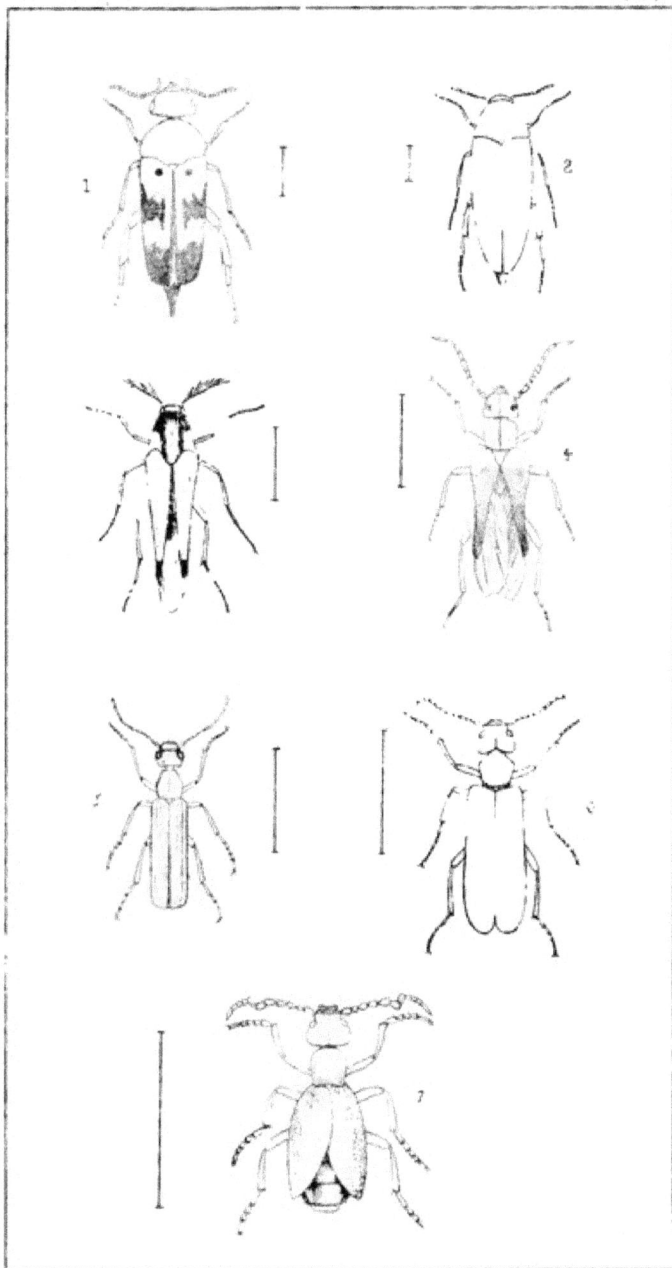

1 MORDELLA   4 SITARIS
2 ANASPIS   5 SYBARIS
3 RHIPIPHORUS   6 CANTHARIS
          7 MELOE

Pl 57.

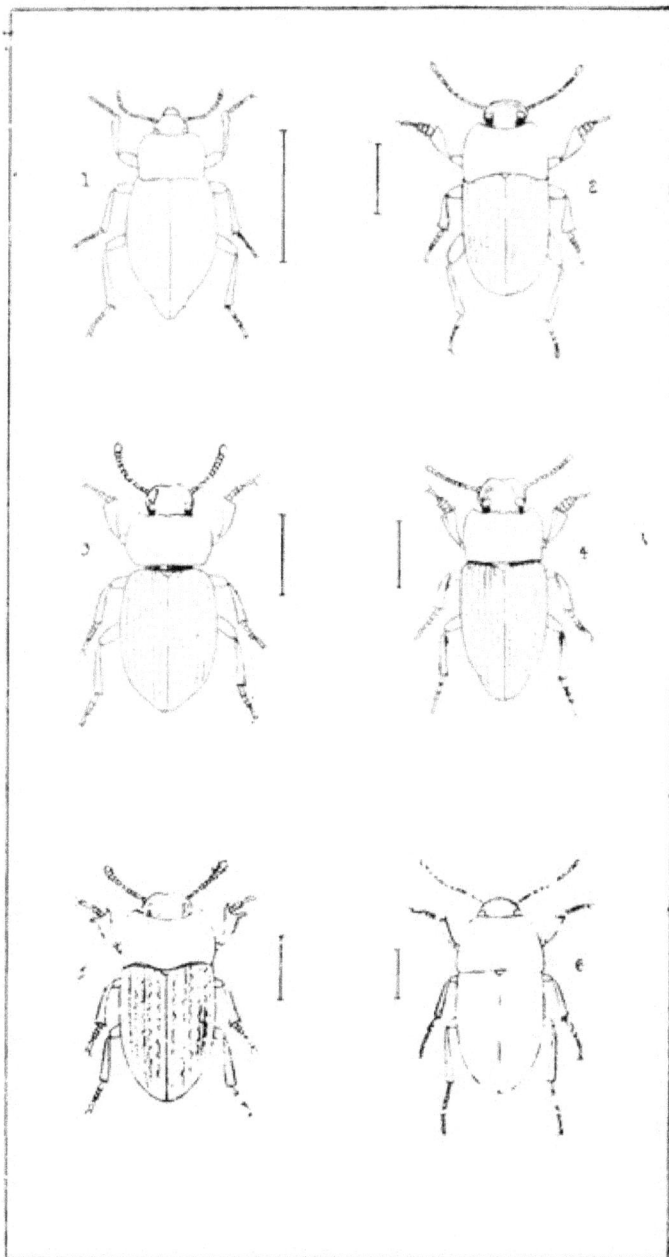

1. BLAPS
2. PEDINUS.
3. HELIOPHILUS
4. PHYLAN
5. OPATRUM.
6. CRYPTICUS

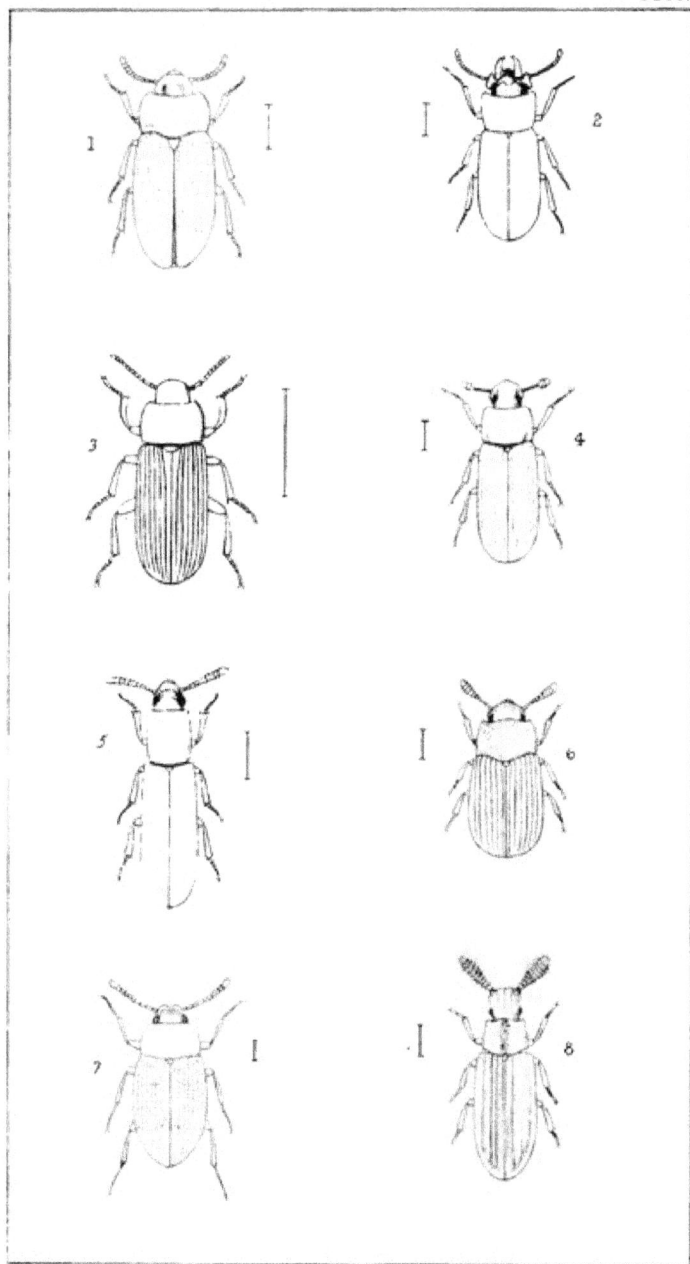

Pl 58.

1. ALPHITOBIUS.
2. ULOMA.
3. TENEBRIO.
4. STENE.

5. HYPOPHLÆUS
6. BOLITOPHAGUS.
7. ALPHITOPHAGUS.
8. SARROTRIUM

Pl 59.

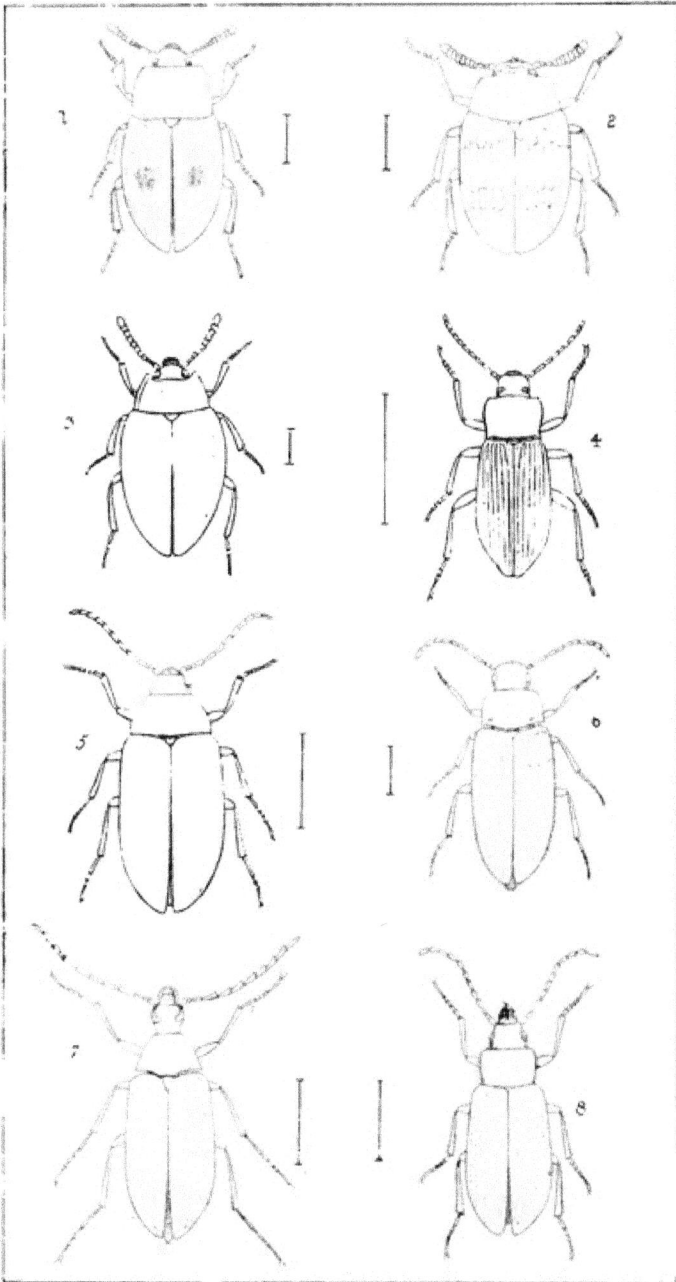

1 PALERIA   5 ERYX
2 DIAPERIS   6 MYCETOCHARIS

Pl. 60

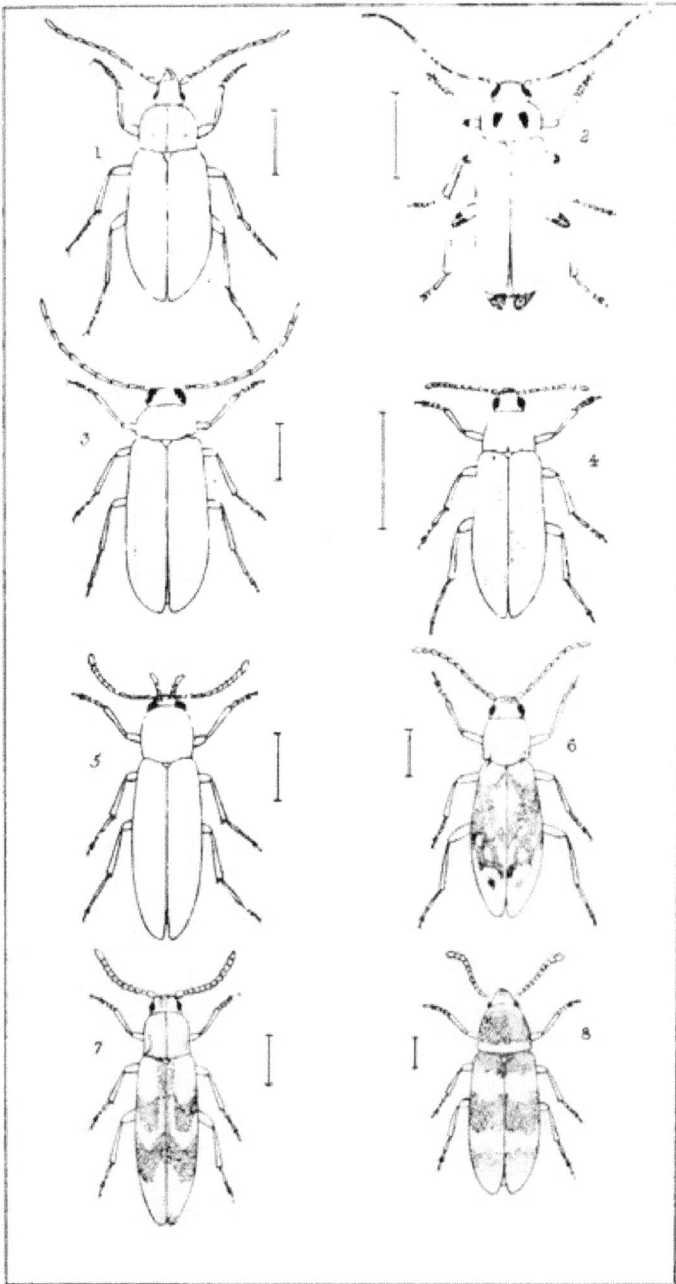

1. CTENIOPUS.
2. NOTHUS
3. CONOPALPUS
4. MELANDRYA
5. PHLOIOTRYA.
6. DIRCÆA.
7. HYPULUS.
8. ABDERA

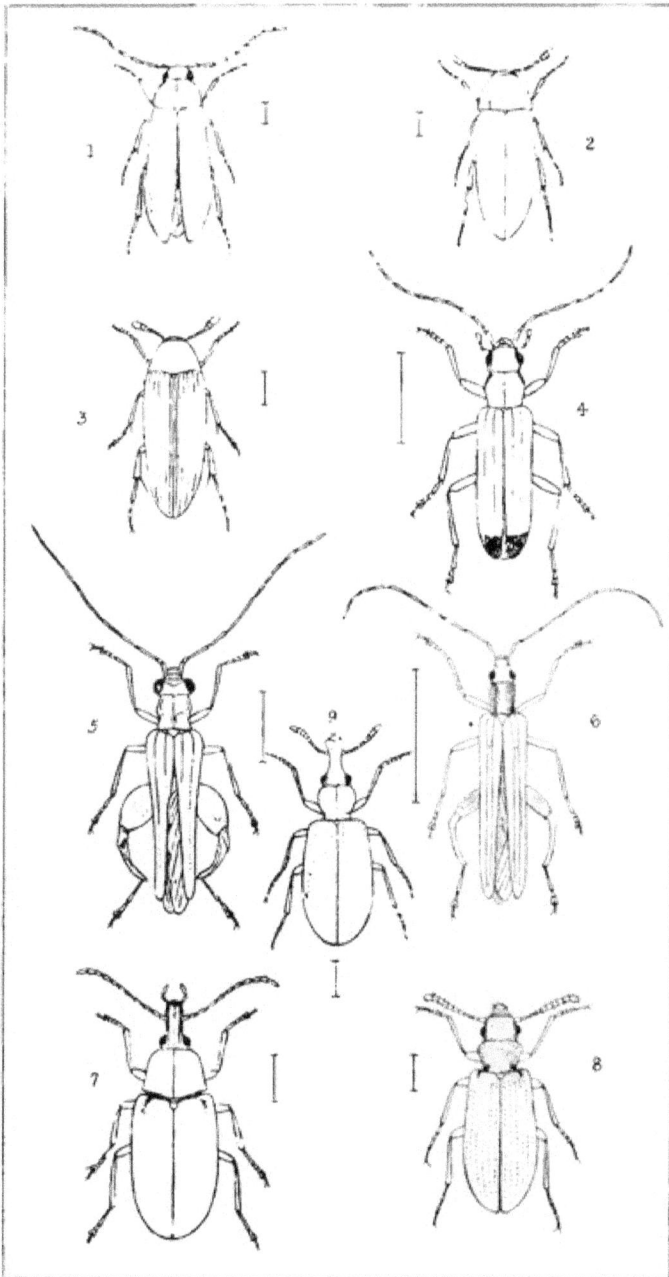

Pl. 61

1. SCRAPTIA
2. HALLOMENUS
3. ORCHESIA
4. ISCHNOMERA
5. ŒDEMERA
6. ONCOMERA
7. MYCTERUS
8. SPHÆRIESTES
9. SALPINGUS.

Pl 62

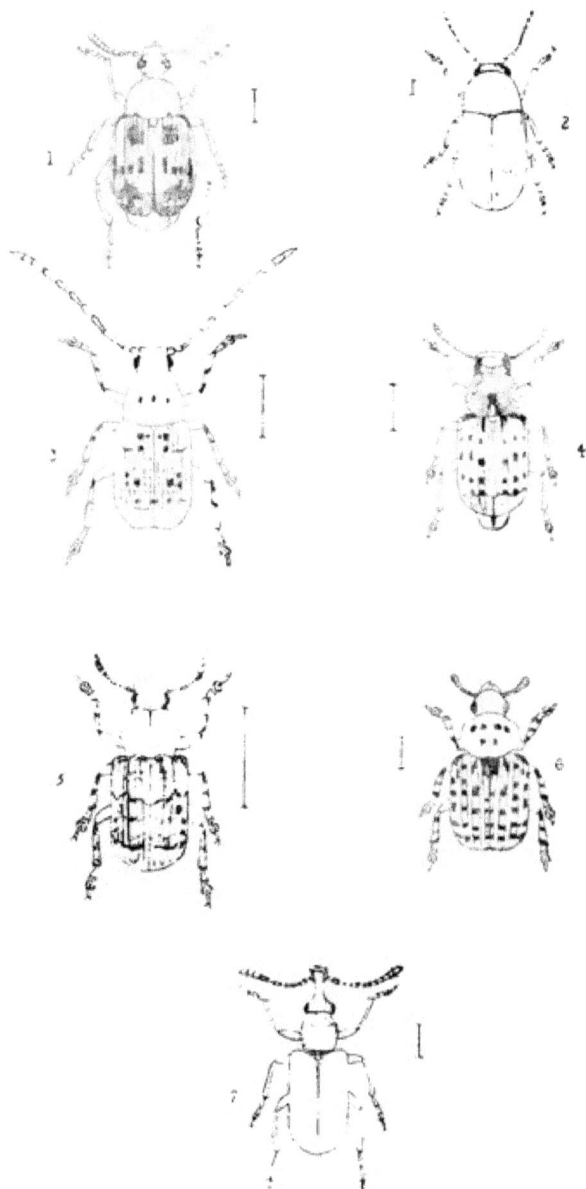

1 BRUCHUS   4 TROPIDERES
2 CHORAGUS   5 PLATYRHINUS
3 ANTHRIBUS   6 BRACHYTARSUS
7 RHINOMACER

Pl 63

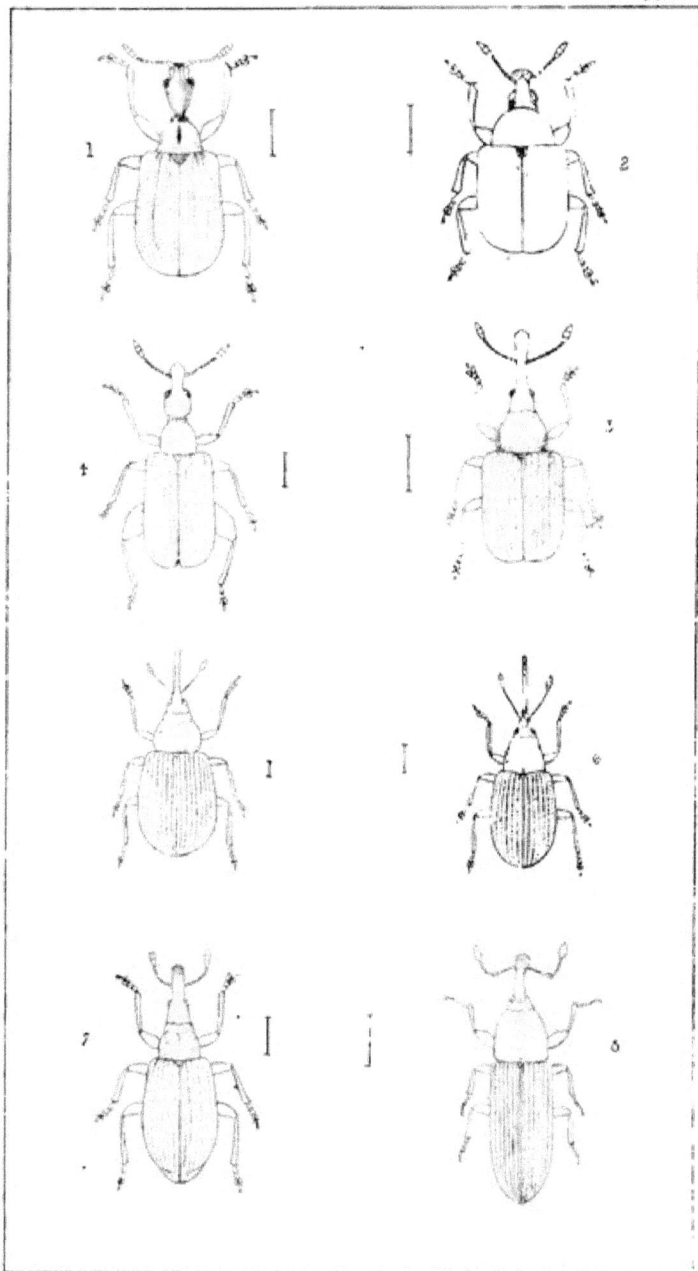

1 APODERUS   5 RAMPHUS
2 ATTELABUS   6 OXYSTOMA

Pl 64

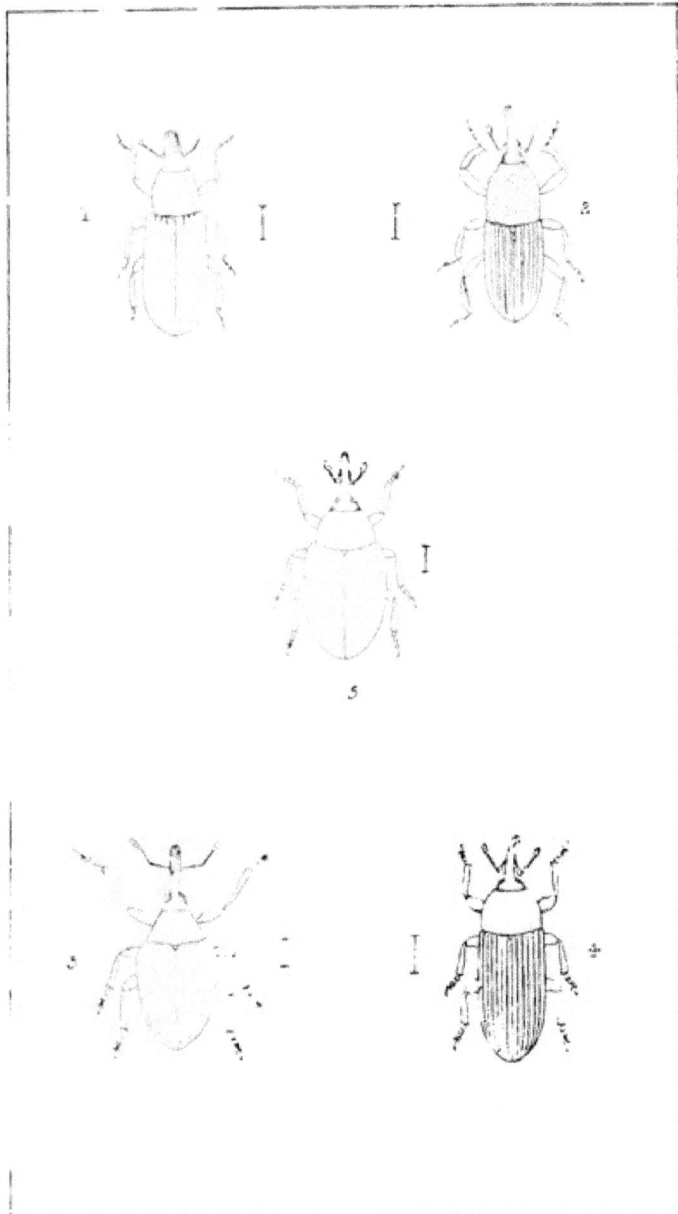

1. RHYNCOLUS   3 SPHÆRULA.
2. CALANDRA   4 MECINUS
5 CYMAETRON.

Pl 65

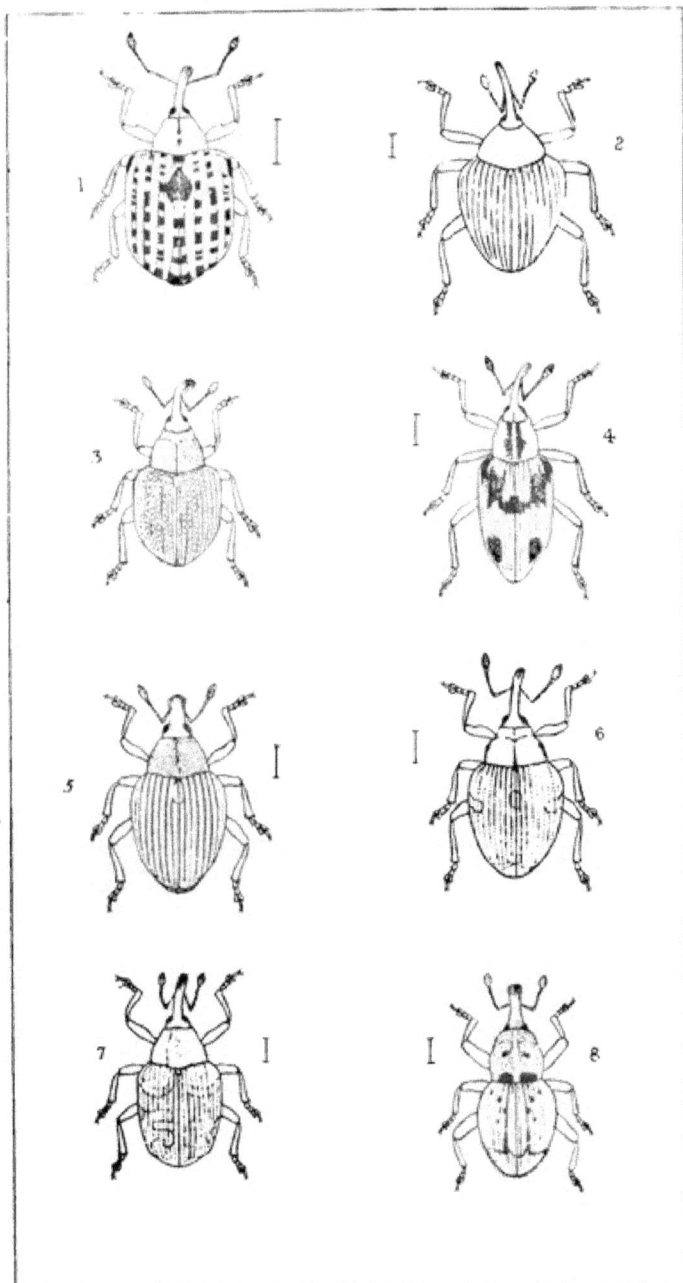

1. CIONUS
2. OROBITIS
3. RUTIDOSOMA
4. POOPHAGUS

5. RHINONCHUS
6. NEDYUS
7. CEUTORHYNCHUS
8. ACALLES

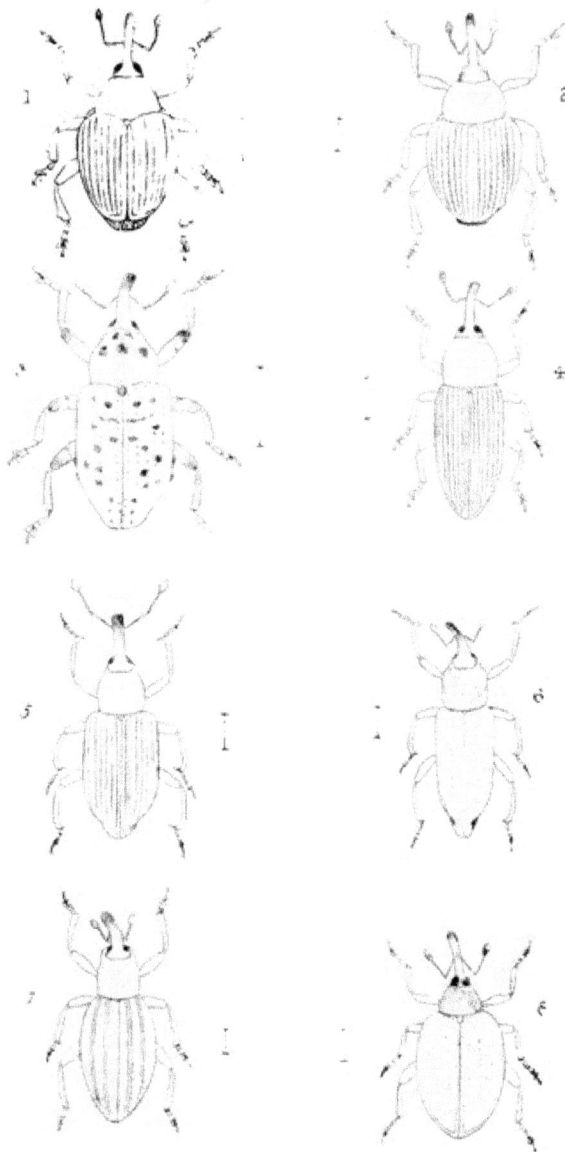

1. MONONYCHUS
2. CÆLIODES
3. CRYPTORHYNCHUS
4. BARIDIS.
5 BAGOUS
6 LYPRUS
7 ORTHOCHÆTES.
8 TACHYERGES.

Pl 67.

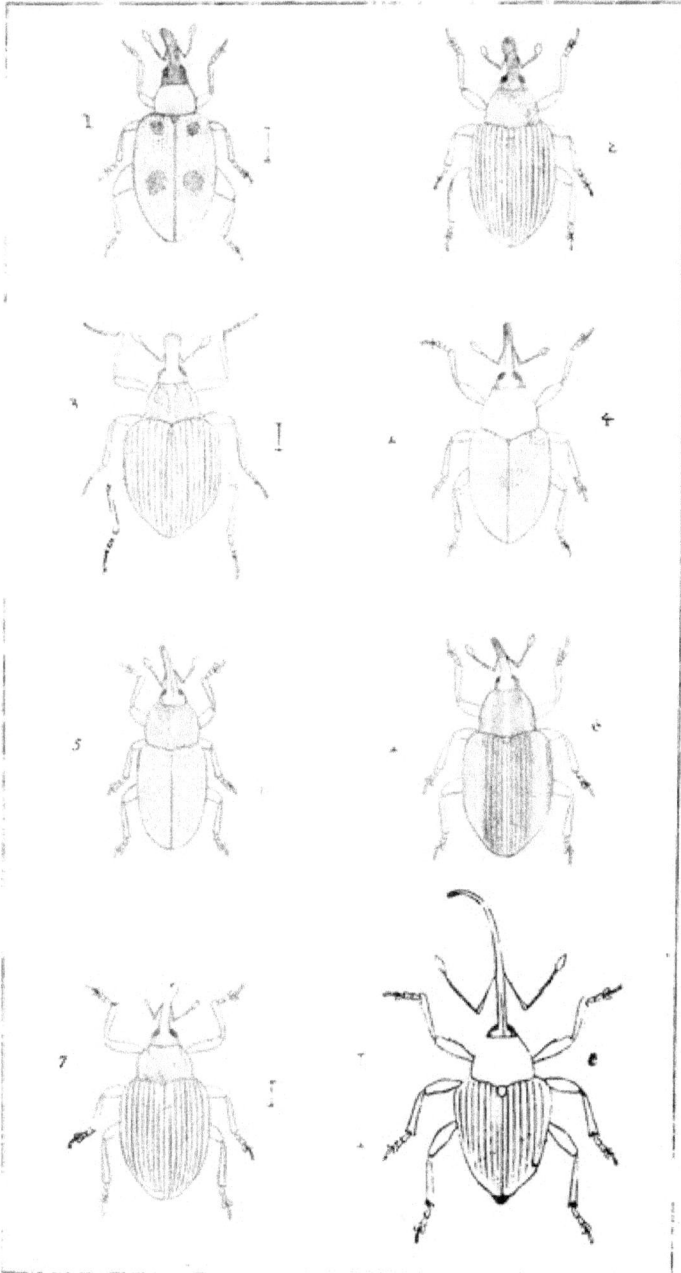

1 ORCHESTES.
2 ANOPLUS.
3. PACHYRHINUS.
4 SIBYNES.

5 MICCOTROGUS.
6 TYCHIUS.
7 AMALUS.
8 BALANINUS

Pl 68.

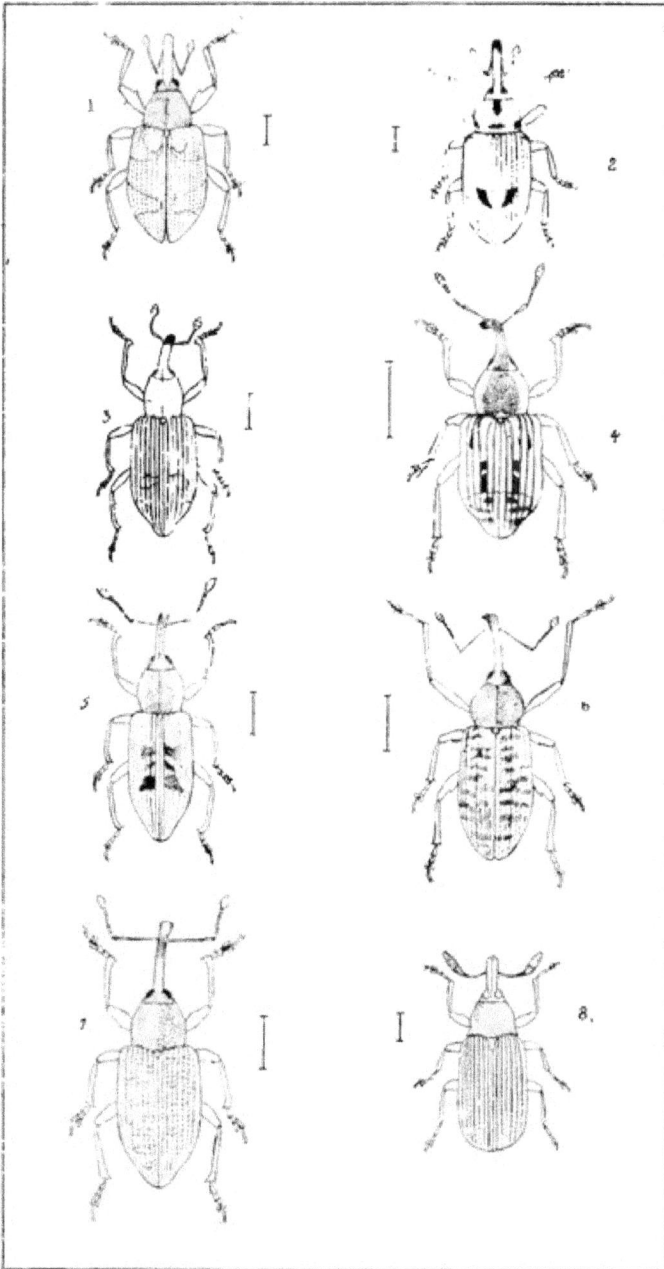

1. ANTHONOMUS.   5. ERIRHINUS
2. ELLESCUS   6. DORYTOMUS
3. HYDRONOMUS   7. NOTARIS
4. GRYPIDIUS.   8. THAMNOPHILUS

Pl 69

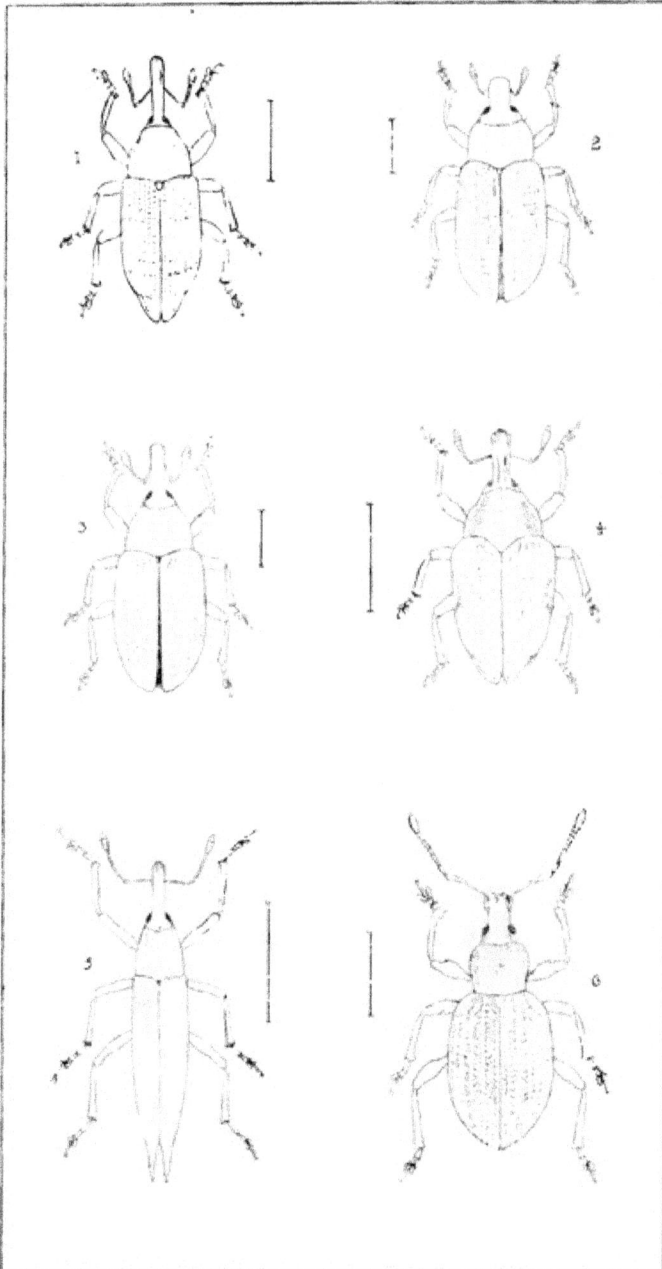

1 PISSODES.
2 RHINOCYLLUS.
3 RHINOBATUS

4 LARINUS
5 LIXUS
6 OTIORHYNCHUS

Pl 70

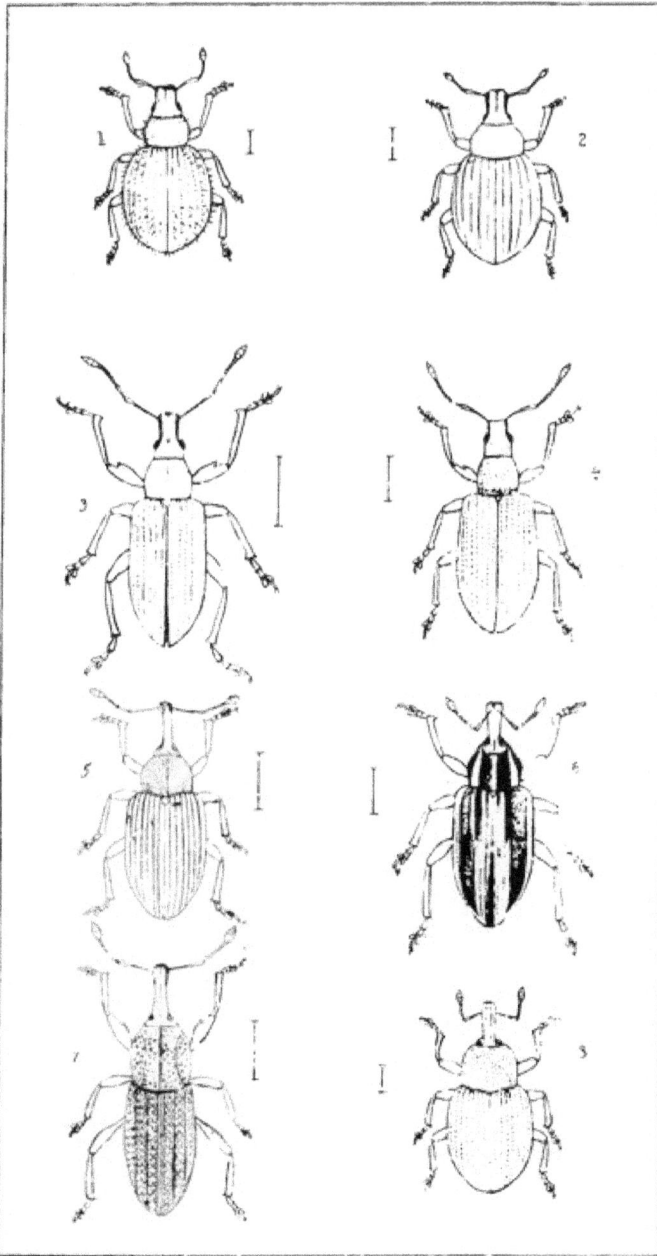

1. OMIAS
2. TRACHYPHLÆAS
3. PHYLLOBIUS.
4. NEMOICUS.

5. PROCAS.
6. PHYTONOMUS
7. PLINTHUS.
8. LEIOSOMA

Pl. 71

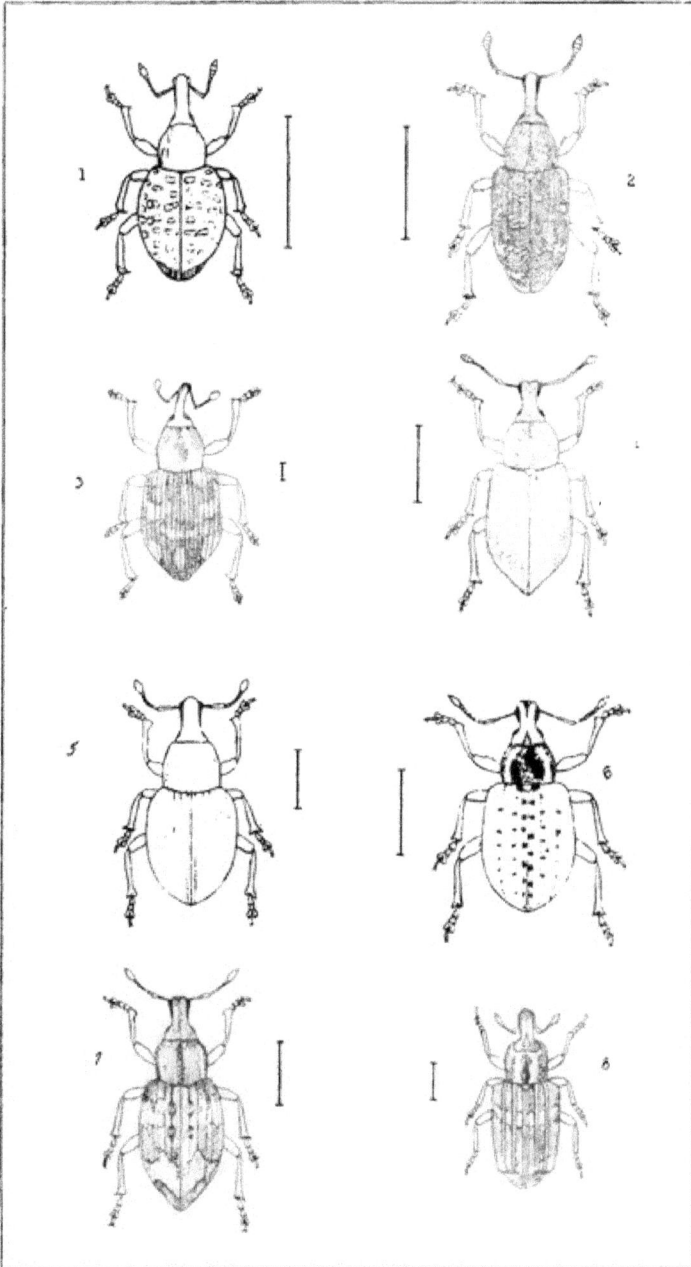

1 MOLYTES
2 HYLOBIUS
3 TANYSPHYRUS
4 MERIONUS

5 BARYNOTUS
6 LIOPHLÆUS
7 ALOPHUS
9 GRONOPS

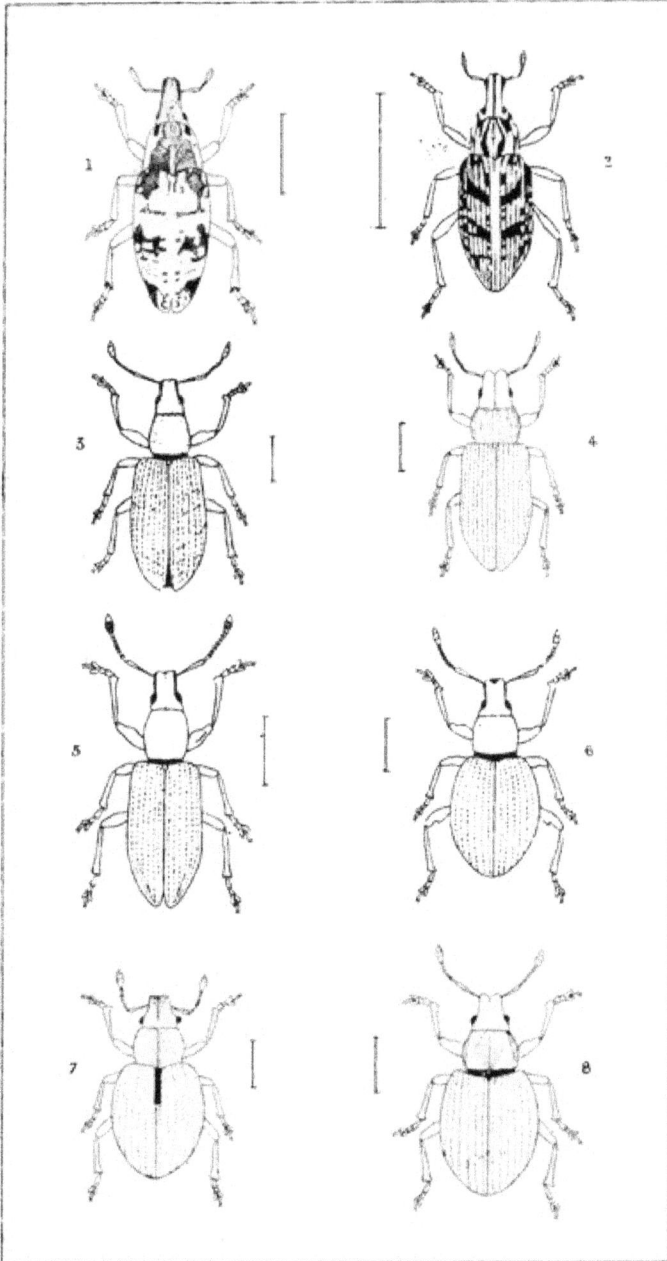

1. BOTHYNODERES
2. CLEONUS
3. POLYDRUSUS
4. SITONA

5. TANYMECUS
6. SCIAPHLUS
7. STROPHOSOMUS
8. ONEORHINUS.

Pl 73

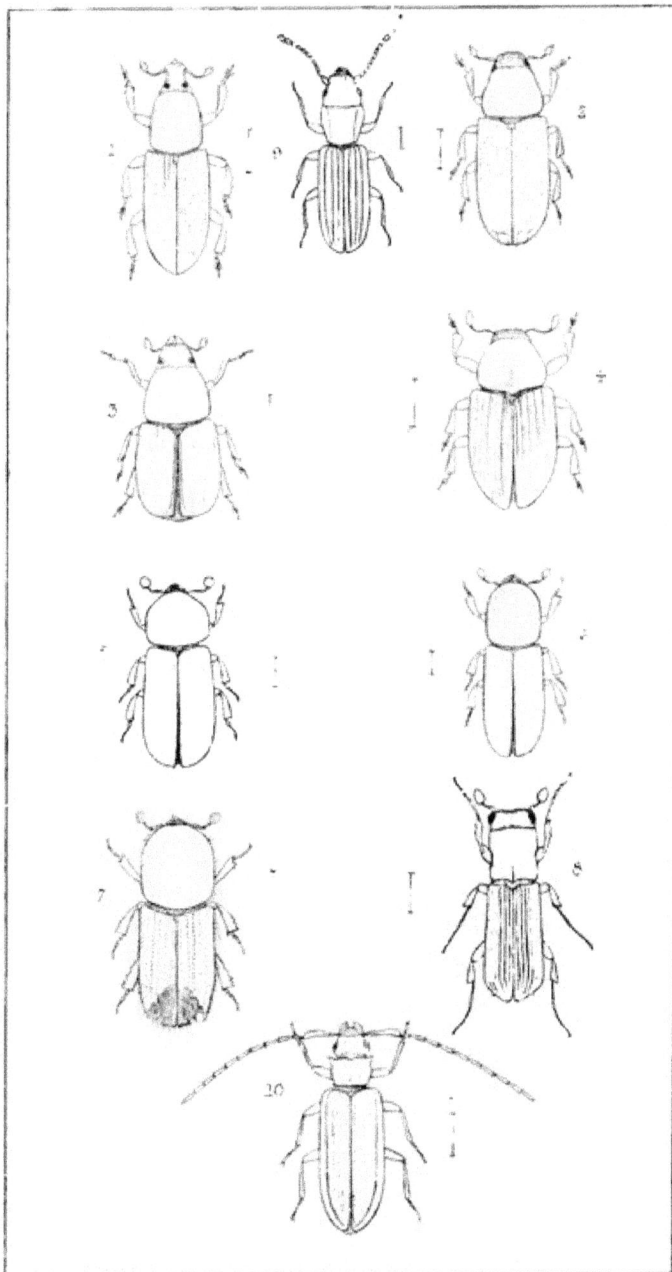

1. HYLASTES
2. DENDROCTONUS
3. SCOLYTUS
4. HYLESINUS
5 TRYPODENDRON

6 POLYGRAPHUS
7. TOMICUS
8 PLATYPUS
9. CUCUJUS
10 ULEIOTA.

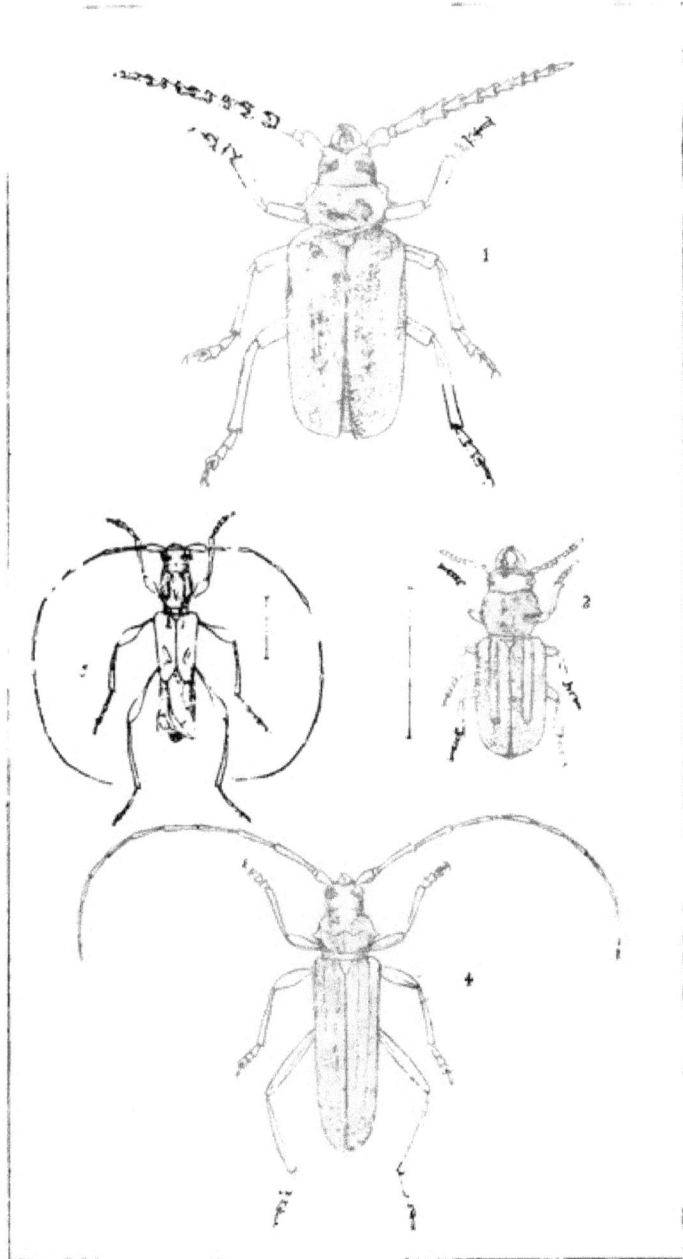

1 PRIONUS   3 NECYDALIS

2 SPONDYLIS   4 AROMIA

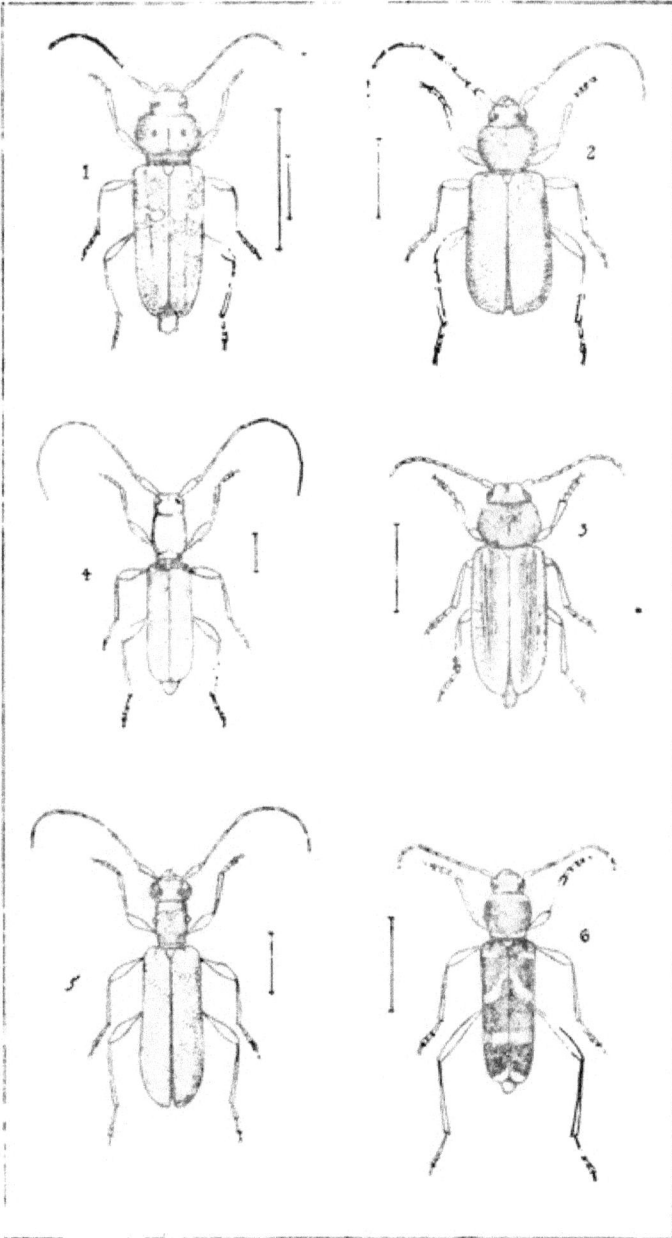

Pl 75

1 HYLOTRUPES
2 CALLIDIUM
3 ASEMUM

4 GRACILIA
5 OBRIUM
6 CLYTUS

Pl 76

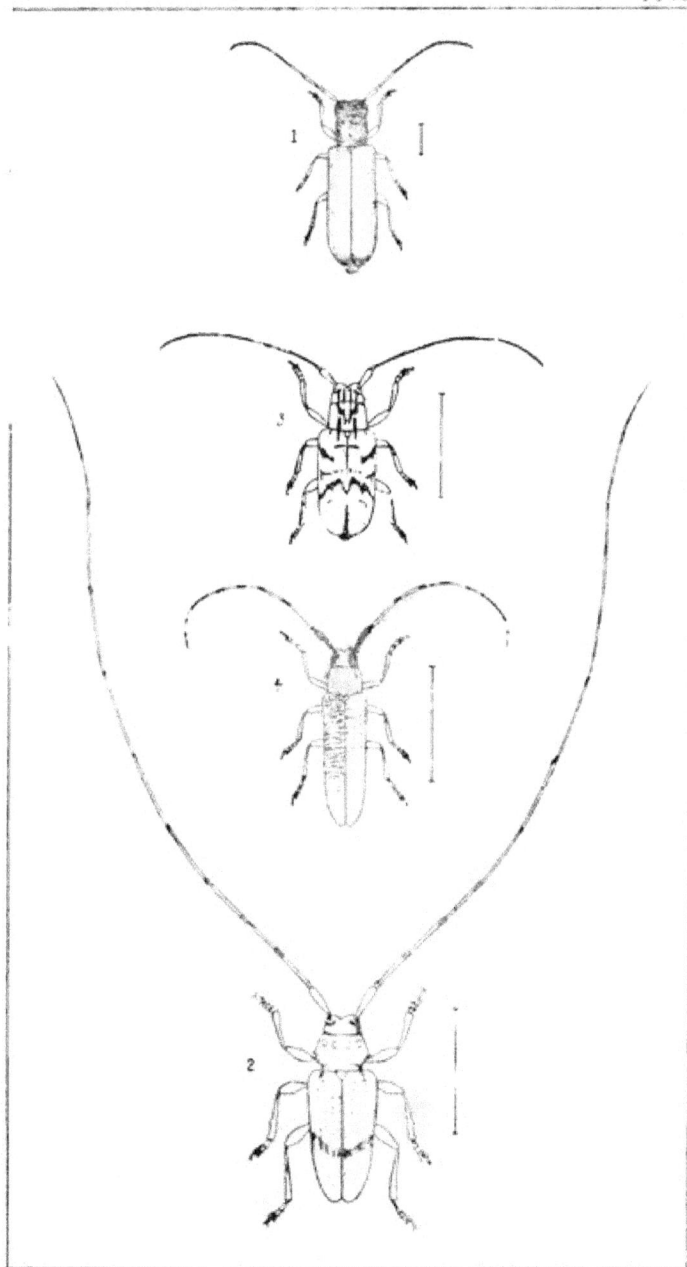

1 TETROPS
2 ACANTHOCINUS
3 APHELOCNEMIA
4 AGAPANTHIA

Pl 27

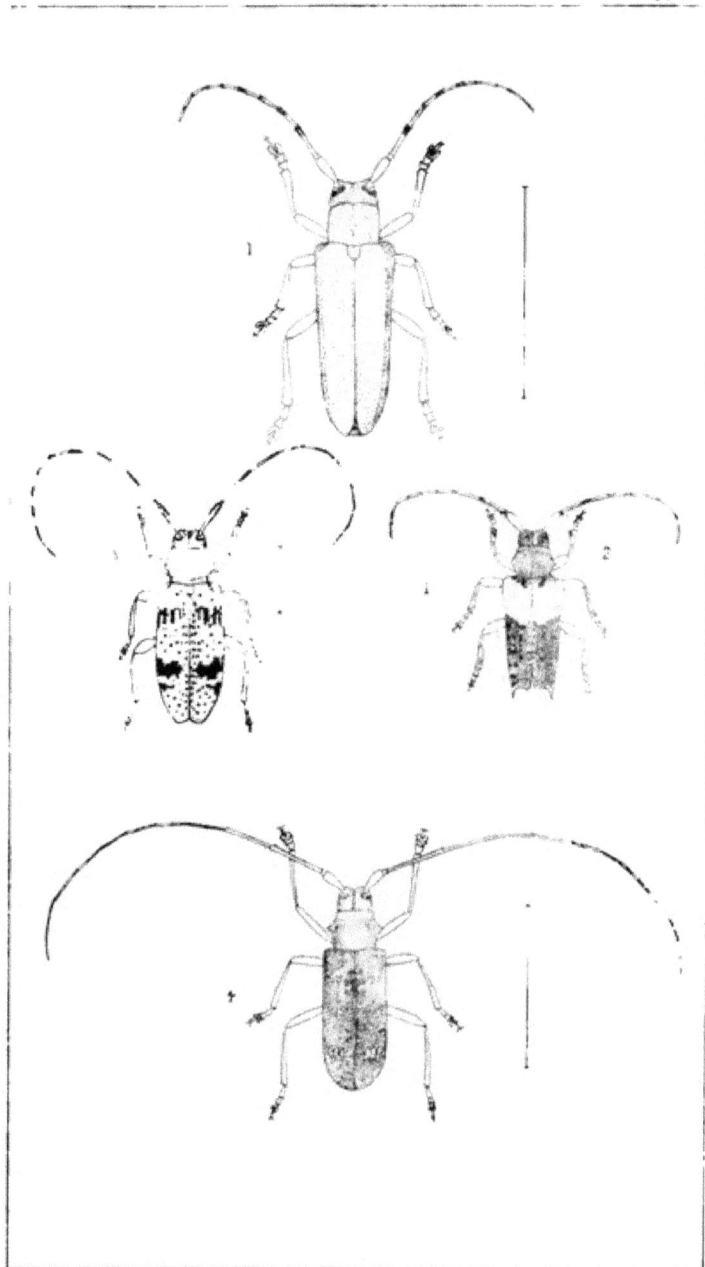

1 SAPERDA   3 LEIOPUS
2 POGONOCERUS.   4 MONOCHAMUS

Pl 78

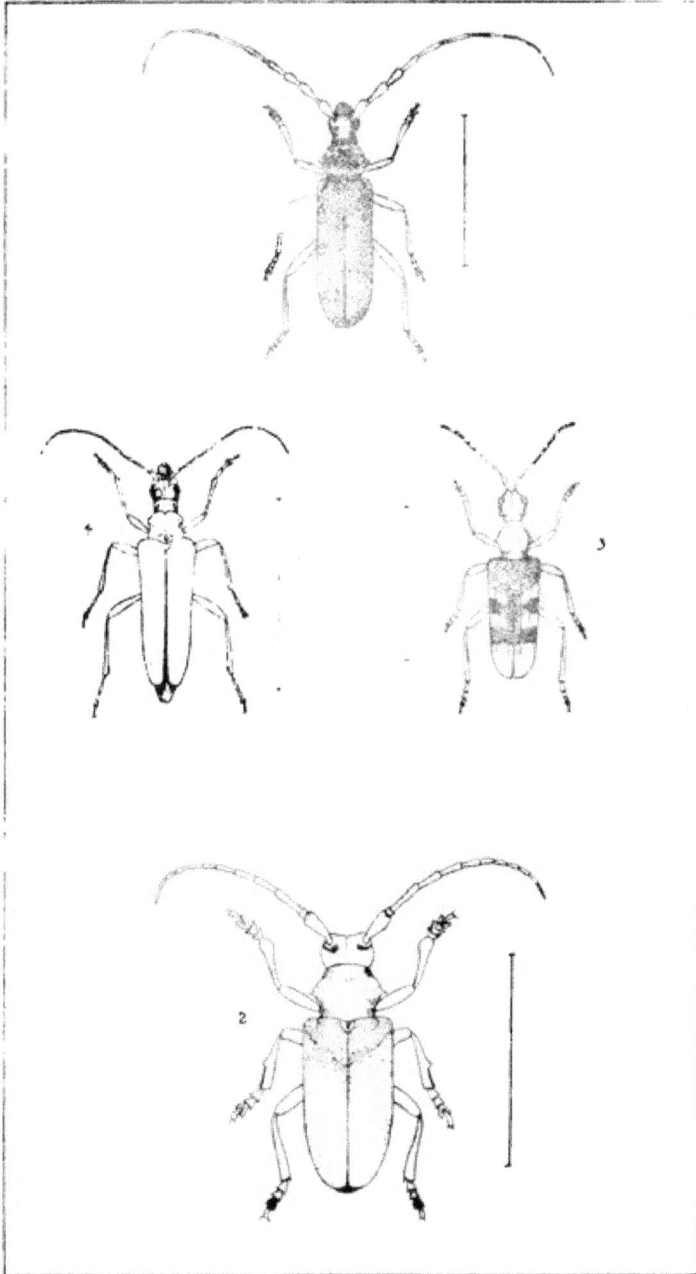

1. CERAMBYX   3 RHAGIUM

2. LAMIA,   4 TOXOTUS

Pl 73

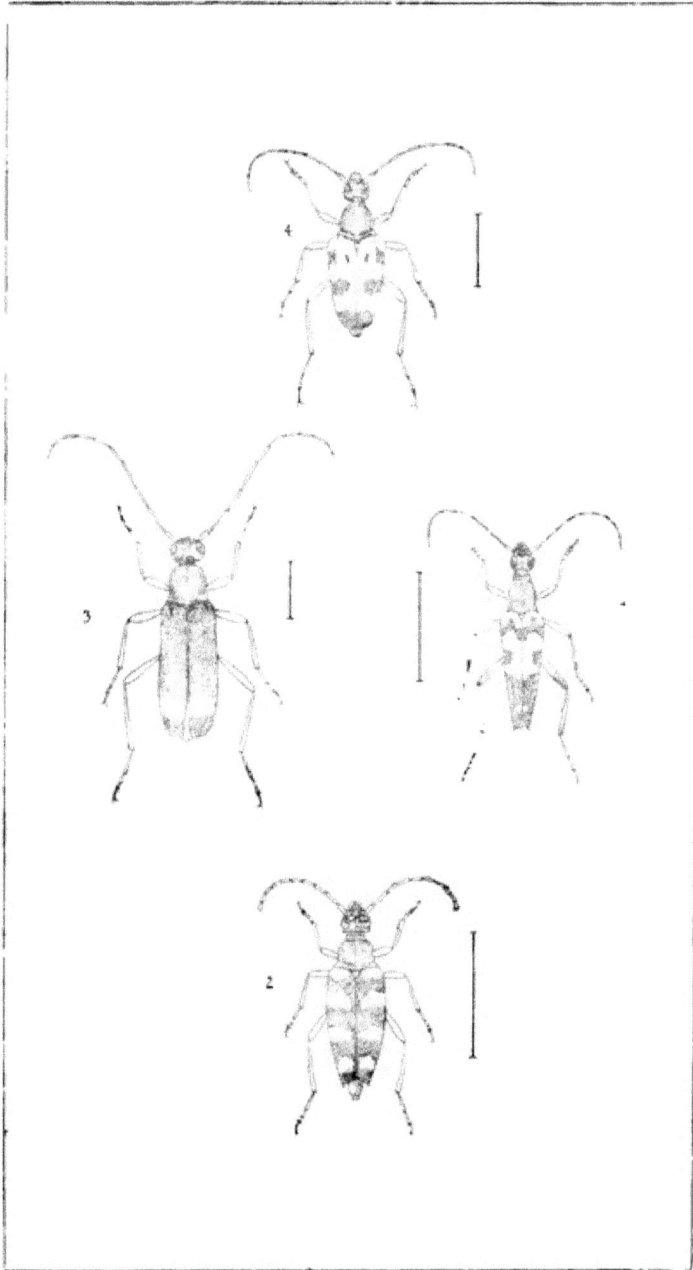

1 STRANGALIA.   3. GRAMMOPTERA
2 LEPTURA   4 PACHYTA

Pl 80

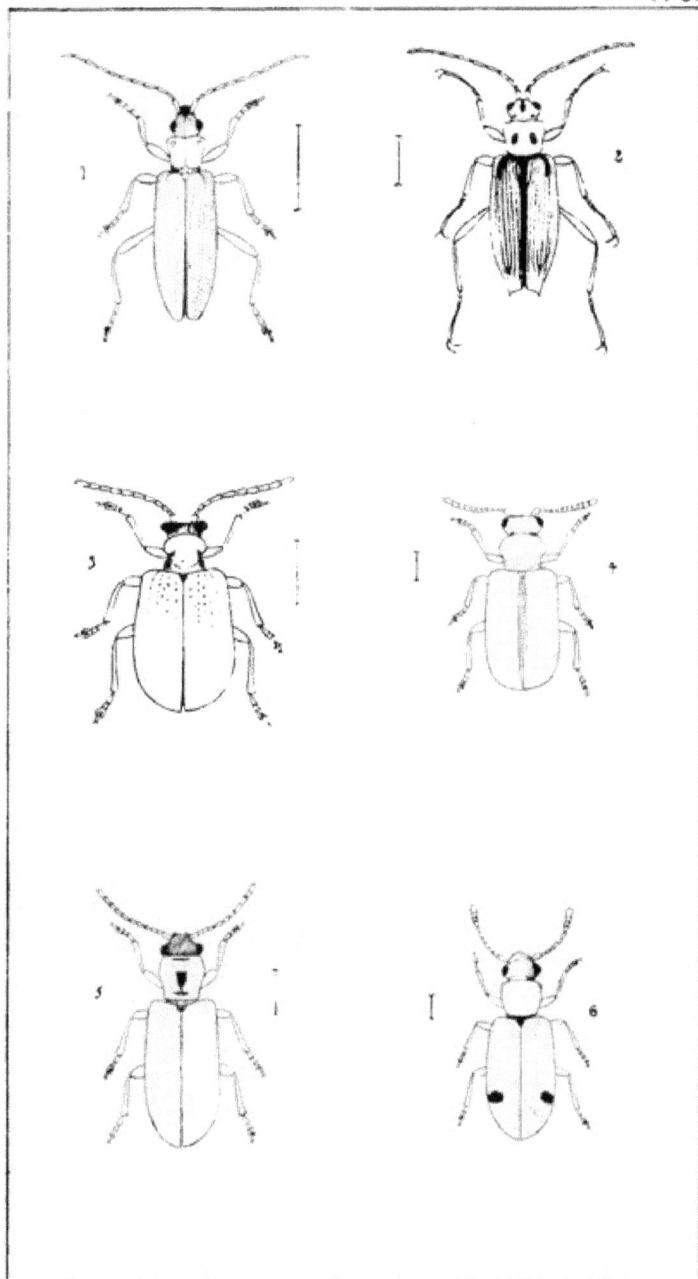

1 DONACIA
2 MACROPLEA
3 CRIOCERIS

4 ZEUGOPHORA
5 ORSODACNA
6 PSAMMÆCHUS

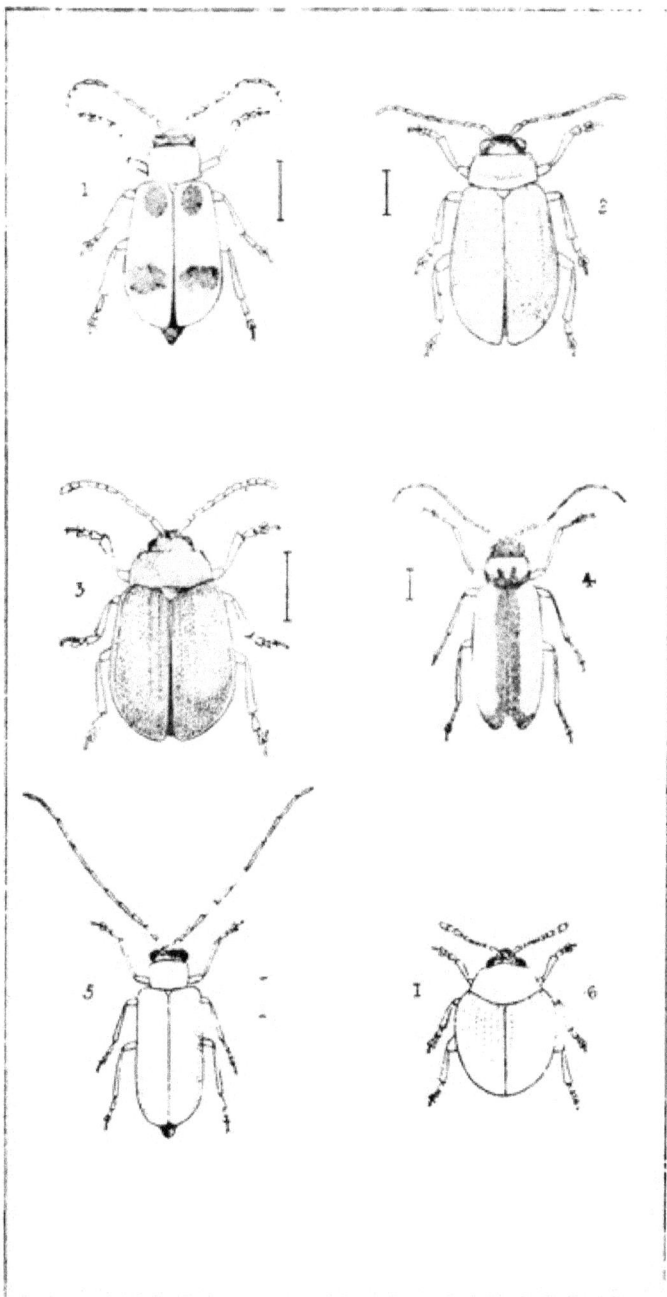

1. AUCHENIA
2. ADIMONIA
3. GALERUCA.
4. CALOMICRUS.
5. LUPERUS
6. MNIOPHILA

Pl 82

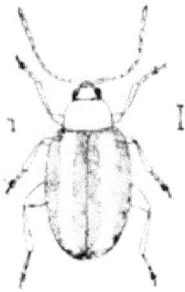

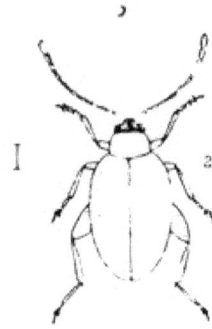

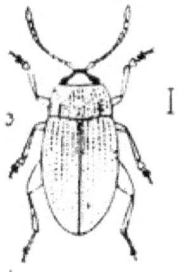

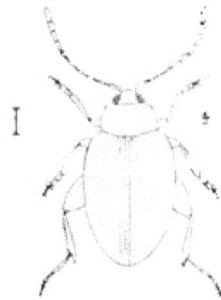

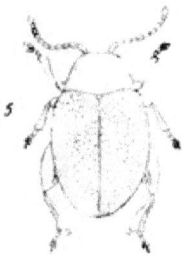

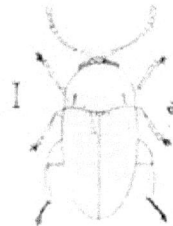

1. HALTICA
2. THYAMIS.
4. MACROCNEMA
5. DIBOLIA

Pl.83

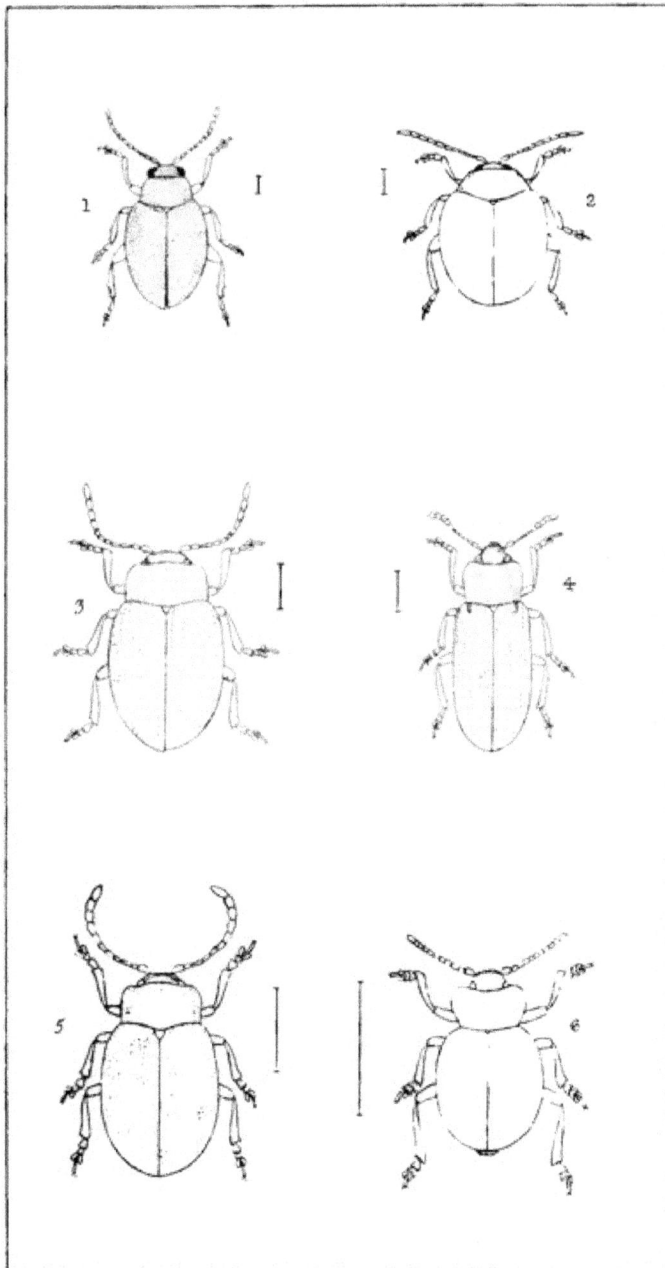

1. CHETOCNEMA.
2 SPHÆRODERMA.
3 PHÆDON.
4. PRASOCURIS
5 CHRYSOMELA.
6 TIMARCHA

Pl 54

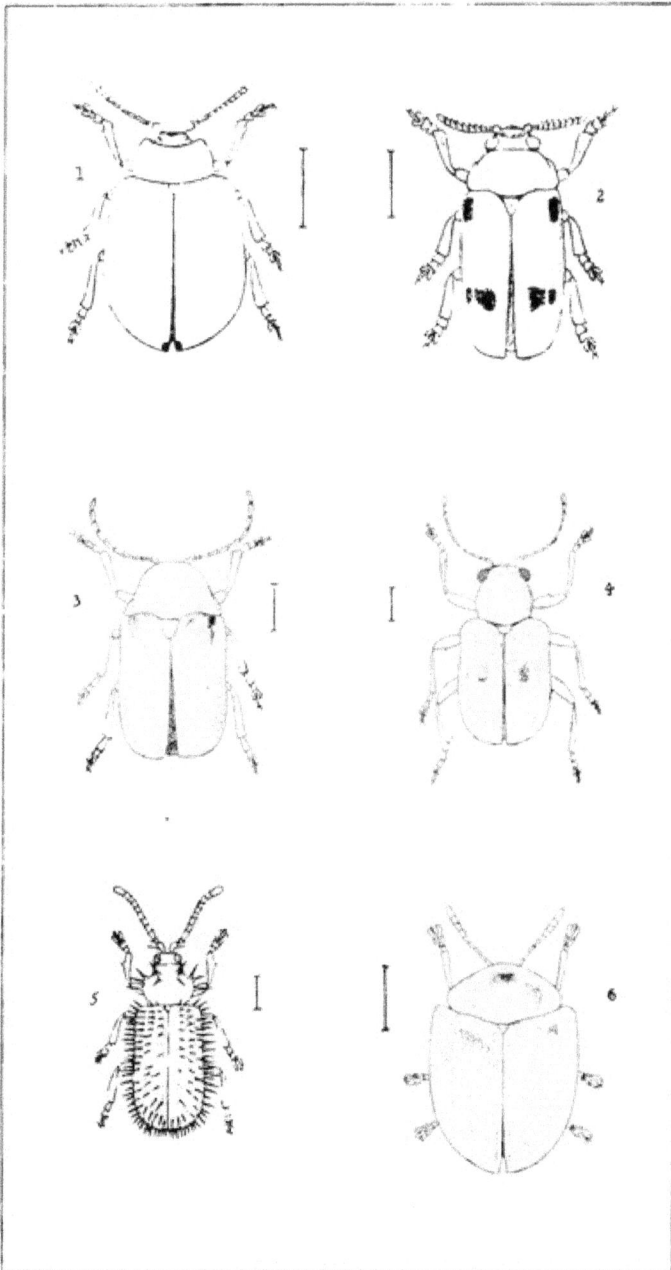

1. MELASOMA
2. CLYTHRA
3. CRYPTOCEPHALUS
4. EUMOLPUS
5. HISPA.
6. CASSIDA

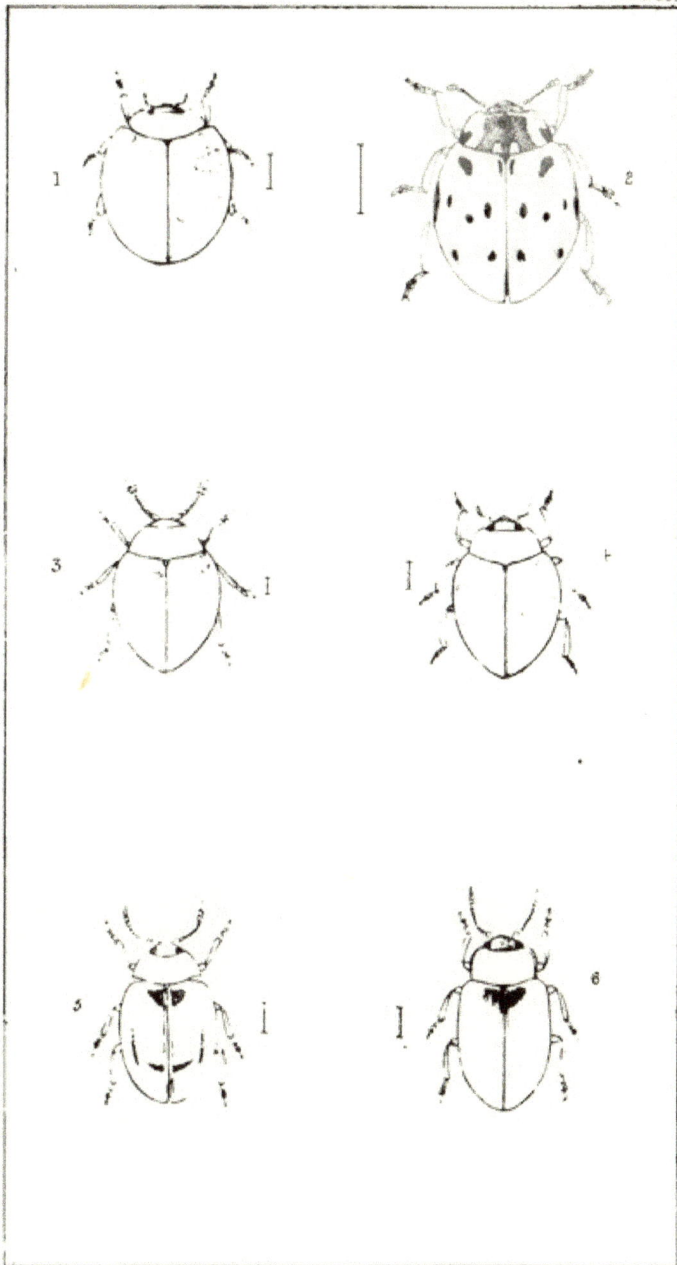

Pl 85

1 CHILOCORUS.   4 SCYMNUS.
2 COCCINELLA   5 RHYZOBIUS.
3 SPHÆROSOMA   6 CACICULA

Pl. 86

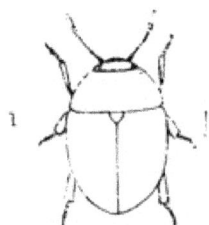

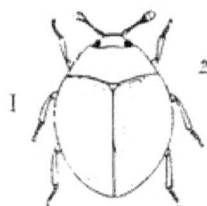

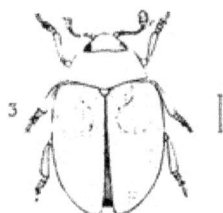

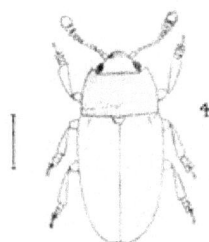

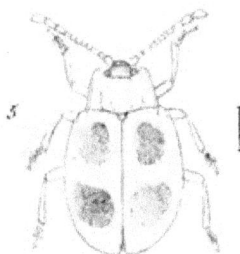

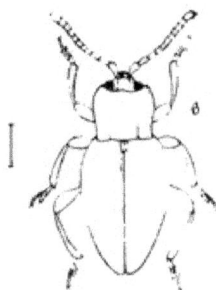

1. PHALACRUS   4. TRIPLAX.
2. ALEXIA   5. ENDOMYCHUS.
3. TRITOMA   6. LYCOPERDINA.

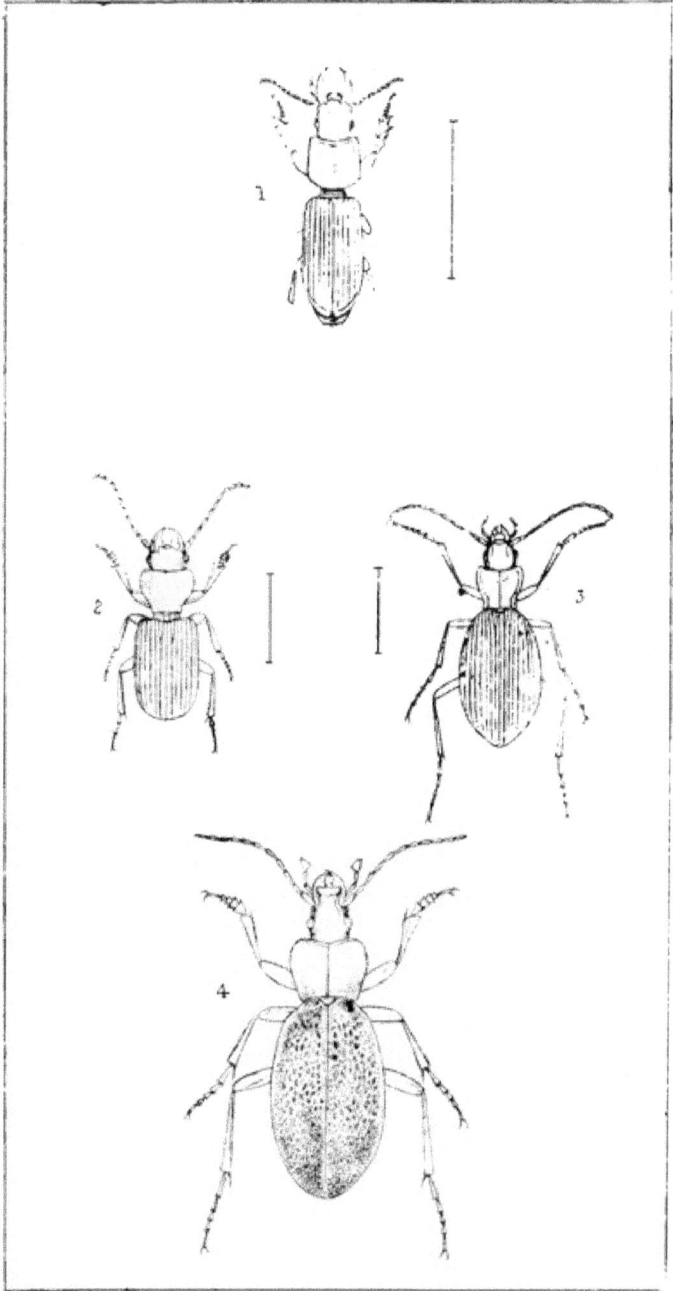

1. OXYSTOMUS.   3. ALPÆUS.
2. DISTOMUS.   4. PROCRUSTES

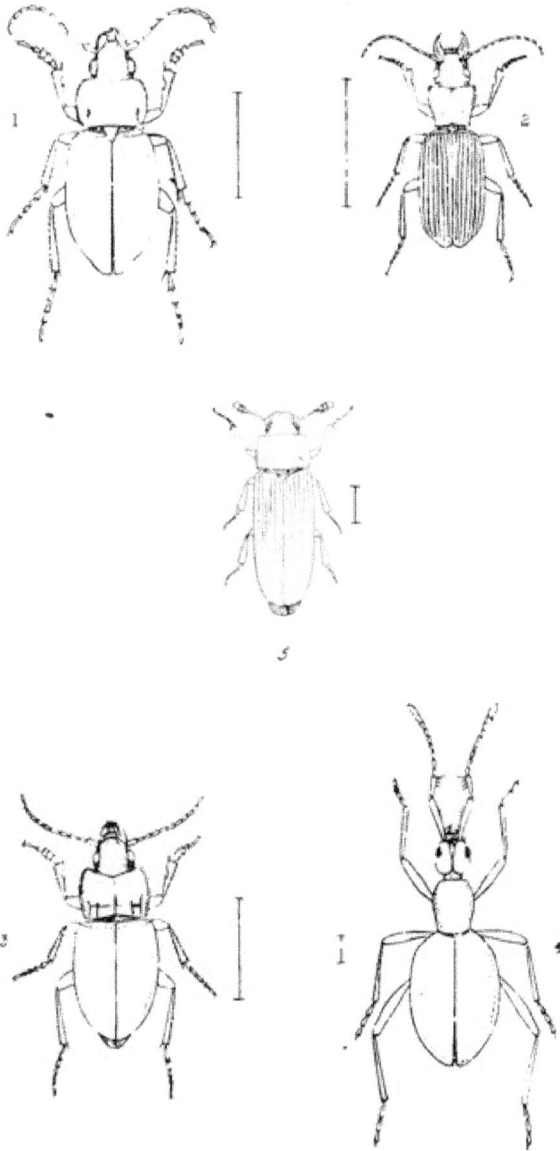

1. SOGINES   3. CHEPORUS.
2. COPHOSUS   4. MASTIGUS.
5. TRIBOLIUM.

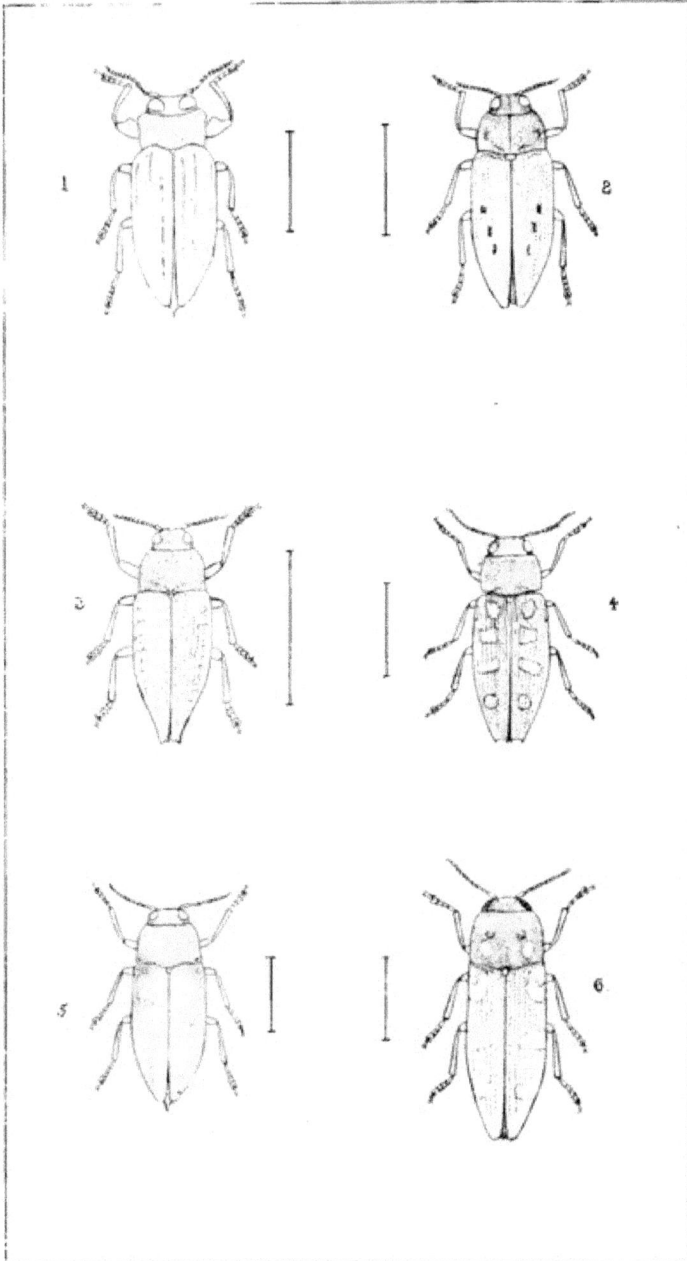

1 CHRYSOBOTHRUS. 4 ANCYLOCHEIRA.
2. LAMPRA   5 MELANOPHILA.
3 DICIRCA   6. PTOSIMA.

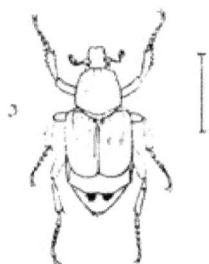

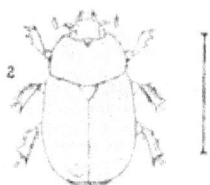

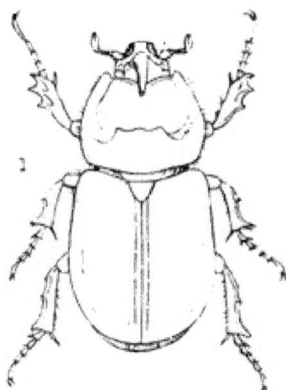

1 ORYCTES   2 DYNASTES

3 VALGUS

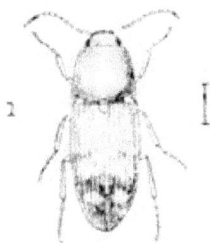

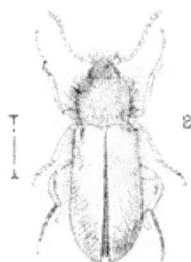

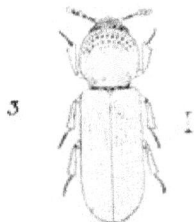

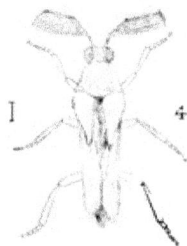

1 DRASTERIUS   3 RHYZOPERTHA
2 ENICOPUS.   4 RHIPIDIUS

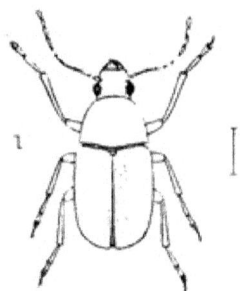

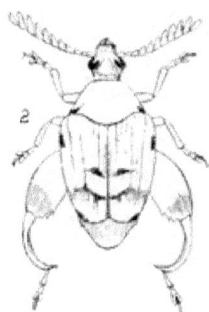

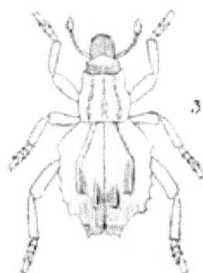

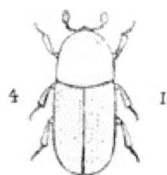

1. PHLŒOBIUS.   3 RHYTIRHINUS
2. CARYOBORUS   4 HYPOTHENEMUS

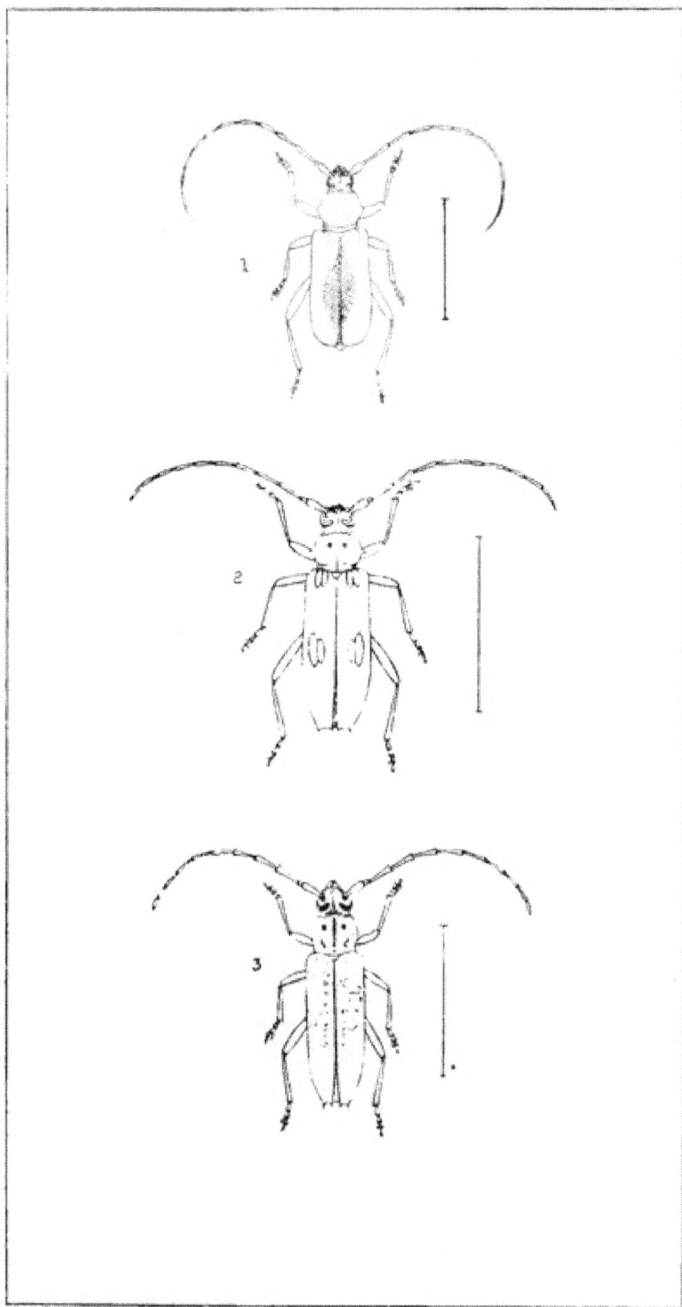

1 PURPURICENUS.   2 EBURIA.
3 ELAPHIDION.

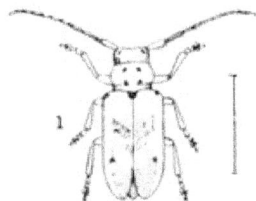

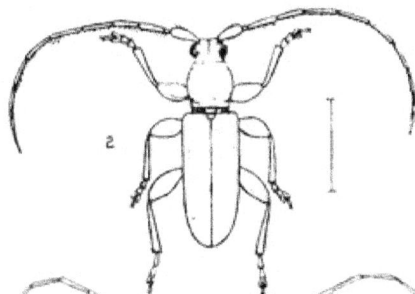

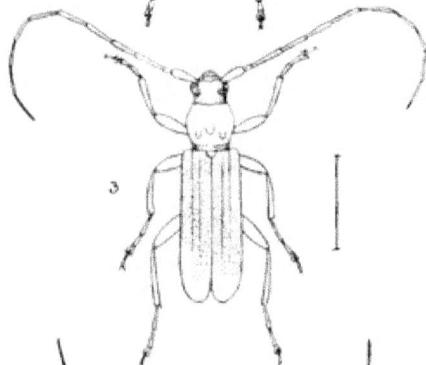

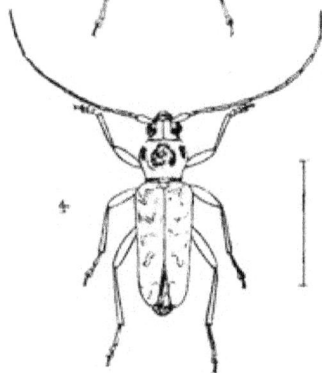

1. TETRAOPES
2. CYLINDERA
3. PENICHROA.
4. AROPALUS.

INDEX.

	Page.	Pl.	Fig.		Page.	Pl.	Fig.
Abax	5	5	3	*Anisocera—see* Anomæocera.			
Abdera	52	60	8	Anisodactylus	3	2	7
Abræus	34	41	8	Anisoplia	37	45	3
Acalles	57	65	8	Anobium	44	53	6
Acanthocinus	66	76	2	Anomala	36	45	2
Achenium	15	18	6	Anomæocera	25	30	2
Acidota	13	15	3	Anommatus	27	33	2
Acilius	12	14	4	Anoplus	58	67	2
Acrodon	5	5	6	Anthaxia	37	46	1
Actephilus	3	3	2	Antherophagus	26	31	1
Adelosia	5	5	1	Anthicus	47	55	6
Aderus	47	55	4	Anthobium	13	15	6
Adimonia	69	81	2	Authonomus	59	68	1
Adrastus	38	46	7	Anthrenus	29	35	7
Ædilis	66	76	2	Anthribus	54	62	3
Aegialia	36	44	1	Apate	44	53	4
Aëpus	6	7	3	Aphanisticus	38	46	3
Agabus	11	13	5	Aphelocnemia	66	76	3
Agapanthia	66	76	4	Aphodius	35	43	6
Agathidium	22	27	5	Apion	55	63	7
Agonum	6	6	8	Aplocnemus	42	51	2
Agrilus	38	46	2	Aploderus	14	16	7
Agriotes	39	46	8	Aplotarsus	40	48	5
Agrypnus	39	47	7	Apoderus	54	63	1
Aleochara	19	23	4	Arcopagus	21	26	2
Alexia	71	85	2	Argutor	4	4	1
Alophus	62	71	7	Arhopalus, *Sup.*	76	8	4
Alpœus, *Sup.*	73	1	3	Aromia	65	74	4
Alphitobius	49	58	1	Asemum	65	75	3
Alphitophagus	50	58	7	Aspidiphorus	29	35	6
Amalus	59	67	7	Asemus	15	18	1
Amara	5	5	4	Astilbus	21	25	2
Amphibolus	32	38	5	Astrapæus	16	20	2
Anaspis	47	56	2	Astynomus	66	76	2
Anchomenus	6	7	2	Atemeles	19	22	8
Ancylocheira, *Sup.*	74	3	4	Athoüs	40	48	7
Anisarthria	26	31	2	Atomaria	26	31	3
Atopa—see Dascillus.				Attagenus	29	35	4

M

	Page.	Pl.	Fig.
Attelabus	55	63	2
Auchenia	69	81	1
Antalia	21	25	4
Badister	7	8	6
Bagoüs	58	66	5
Balaninus	59	67	8
Baris—see Baridius.			
Beridius	57	66	4
Barynotus	62	71	5
Batrisus	21	25	8
Bembidium	9	11	1
Berosus	32	39	3
Biphyllus	26	32	4
Bisnius	16	19	5
Bitoma	27	33	7
Blaps	48	57	1
Bledius	14	17	4
Blemus	6	7	5
Blethisa	8	10	5
Bolbocerus	35	43	3
Bolitobius	18	21	8
Bolitochara	20	24	3
Bolitophagus	49	58	6
Bostrichus	44	53	3
Bothynoderes	62	72	1
Brachinus	2	1	5
Brachysomus— see Omias.			
Brachytarsus	54	62	6
Bradycellus	7	7	6
Bradytus	5	5	7
Broscus	4	4	4
Bruchus	54	62	1
Bryaxis	21	26	4
Byrrhus	30	36	5
Bythinus	26	26	3
Byturus	27	32	7
Cacicula	72	86	6
Caclioides	57	66	2
Cafius	16	19	4
Calandra	56	64	2
Calathus	5	6	2
Callicerus	20	23	6
Callidium	65	75	2
Culomicrus	69	81	4
Calodera	20	24	5
Callistus	7	8	1
Calosoma	8	10	1
Campta	25	29	7
Campylus	40	48	8
Cantharis	48	56	6
Carabus	8	10	2
Cardiapus	70	82	6
Cardiophorus	40	48	4
Carpophilus	25	30	6

	Page.	Pl.	Fig.
Caryoborus, *Sup.*	75	6	2
Cassida	68	84	6
Cateretes	25	30	3
Catops	23	28	3
Celia	5	5	5
Centroglossa	19	22	4
Cerambyx	65	78	1
Cercyon	33	40	6
Cerylon	27	32	8
Cetonia	37	45	7
Centorhynchus	57	65	7
Chætarthria	33	40	3
Chætocnema	70	83	1
Cheporus, *Sup.*	73	2	3
Chilocorus	72	86	1
Chlænius	7	8	3
Choleva	23	28	4
Choragus	54	62	2
Chrysobothris, *Sup.*	74	3	1
Chrysomela	70	83	5
Cicindela	1	1	1
Cicones	27	33	6
Cillenum	9	12	2
Cionus	56	65	1
Cis	44	53	2
Cistela	51	59	7
Clambus	22	27	4
Claviger	21	25	5
Cleonus	62	72	2
Cleopus—see Cionus.			
Clerus	43	52	5
Clivina	3	2	4
Clypeaster	22	27	3
Clythra	71	84	2
Clytus	65	75	6
Cnemidotus	10	12	5
Cneorhinus	63	72	8
Coccinella	72	86	2
Colon	23	28	1
Colydium	28	34	5
Colymbetes	11	14	1
Conopalpus	51	60	3
Conurus	18	21	6
Cophosus, *Sup.*	73	2	2
Copris	35	43	4
Coprophilus	13	16	3
Corticaria	26	31	8
Corylophus—see Clypeaster.			
Corynetes	44	52	7
Coryphium	13	15	2
Cossonus	55	63	8
Creophilus	17	21	2
Crioceris	68	80	3
Cryptarcha	24	29	5
Crypticus	49	57	6
Cryptobium	15	18	7
Cryptocephalus	71	84	3

	Page.	Pl.	Fig.		Page.	Pl.	Fig.
Cryptohypnus .	39	47	9	Ellescus . .	59	68	2
Cryptophagus .	26	30	9	Elmis . .	31	37	4
Cryptorhynchus .	57	66	3	Elodes . . .	41	49	3
Ctenicerus . .	40	48	2	Emus . .	17	21	1
Cteniopus . .	51	60	1	Encephalus . .	19	23	1
Ctenonychus—see Synaptus.				Endomychus . .	72	85	5
Cnenjus . .	65	73	9	Engis . .	26	31	4
Curtomerus—see Cylindera.				Enicocerus . .	31	38	3
Curtonotus . .	5	5	8	Enicopus, Sup. .	75	5	2
Cybister . .	12	14	5	Epaphius . .	6	7	4
Cychrus . .	8	10	3	Ephistemus . .	30	36	8
Cyclonotum .	33	40	4	Epomis . .	7	8	4
Cylindera, Sup. .	76	8	3	Erirhinus . .	59	68	5
Cymindis . .	2	1	4	Eryx . .	50	59	5
Cypha . .	18	21	5	Evæsthetus . .	15	18	3
Cyphon—see Elodes.				Eubria . .	41	49	4
				Euplectus . .	21	25	6
Dascillus . .	41	49	1	Euglenes—see Aderus.			
Dabytes . .	43	51	3	Eumolpus . .	71	84	4
Deinopsis .	19	22	3	Euryporus . .	17	20	6
Demetrias . . .	2	1	7	Euthela . .	22	26	8
Dendroctonus .	63	73	2				
Dendrophilus .	33	41	3	Falagria . .	21	25	3
Deporaüs .	55	63	5				
Dermestes . .	29	35	2	Gabrius . .	16	19	3
Diachromus .	3	2	8	Galeruca . .	69	81	3
Dianoüs . .	14	17	6	Georyssus . .	31	37	5
Diaperis . .	50	59	2	Geotrupes . .	35	43	1
Dibolia . .	70	82	5	Gibbium . .	46	54	8
Dicerca, Sup. .	74	3	3	Gnorimus . .	37	45	6
Diglossa . .	19	22	6	Goerius . .	17	20	7
Dinarda . .	19	22	7	Gracilia . .	65	75	4
Dinoderus . .	44	53	5	Grammoptera .	67	79	3
Dircæa .	52	60	6	Gronops . .	62	71	6
Distomus, Sup. .	73	1	2	Grypidius . .	59	68	4
Dolichosoma .	43	51	4	Gymnætron . .	56	64	5
Dolopius . .	39	47	1	Gyrinus . .	12	13	7
Donacia . .	68	80	1	Gyrohypnus . .	16	19	1
Dorcatoma . .	45	53	8	Gyrophæna . .	19	23	2
Dorcus . .	34	42	2				
Dorgtonus . .	59	68	5	Haliplus . .	10	12	4
Drasterius, Sup. .	74	5	1	Hallomenus . .	52	61	2
Drilus . .	41	49	7, 8	Haltica . .	69	82	1
Dromius . .	2	1	8	Harpalus . .	3	3	1
Dryophilus . .	44	53	7	Hedobia . .	45	54	5
Dryops . .	31	37	3	Heliophilus . .	48	57	3
Drypta . .	1	1	2	Helobia . .	8	9	4
Dyctiopterus .	42	50	1	Helodes—see Prasocuris.			
Dynastes, Sup. .	74	4	2	Helophorus . .	31	39	1
Dyschirius . .	3	2	5	Helops . .	50	59	4
Dytiscus . .	11	14	2	Hesperophilus .	14	17	3
				Heterocerus . .	30	37	1
Eburia, Sup. .	75	7	2	Heterothops . .	16	19	6
Ectinus . .	39	47	3	Hispa . .	68	84	5
Elaphidion, Sup. .	76	7	3	Hister . .	33	41	2
Elaphrus .	8	10	4	Holoparamecus .	26	32	1
Elater . .	39	47	5	Homalota . .	20	23	7

	Page.	Pl.	Fig.		Page.	Pl.	Fig.
Hoplia	37	45	4	Ludius	40	48	1
Hydaticus	12	14	3	Luperus	69	81	5
Hydræna	33	38	6	Lycoperdina	72	85	6
Hydrobius	32	40	1	Lyctus	28	34	8
Hydrochus	31	38	2	Lymexylon	46	51	5
Hydronomus	59	68	3	Lymnæum	9	12	3
Hydrophilus	32	39	5	Lyprus	58	66	6
Hydroporus	11	13	2				
Hydroüs	32	39	4	Macronema	70	82	4
Hygronoma	20	23	8	Macropleä	68	80	2
Hygrotus	11	13	1	*Magdalis—see* Thamnophilus.			
Hylastes	63	73	1	Malachius	42	51	1
Hylecætus	46	51	6	Malthinus	42	50	6
Hylesinus	63	73	4	Mautura	69	82	3
Hylobius	61	71	2	Masoreus	3	3	5
Hylotrupes	65	75	1	Mastigus, *Sup.*	73	2	4
Hypera—see Phytonomus.				Mecinus	56	64	4
Hyphidrus	10	12	7	Medon	15	18	4
Hypocyphthus—see Cypha.				Megaeronus	18	22	1
Hypolithus	39	47	8	Megaladerus	22	26	7
Hypophlæus	49	58	5	Megatoma	29	35	5
Hypothenemus, *Sup.*	75	6	4	Megarthrus	13	16	1
Hypulus	52	60	7	Melandrya	51	60	4
				Melanophila, *Sup.*	74	3	5
Ilybius	11	13	6	Melanotus	40	47	10
Ips	25	30	7	Melasis	38	46	5
Ischnomera	52	61	4	Melasoma	70	84	1
Ischnopoda	20	25	1	Meligethes	25	29	8
				Meloë	48	56	7
Laccobius	32	39	2	Melolontha	36	44	6
Laccophilus	11	13	4	Merionus	62	71	4
Lagria	46	55	2	Mesosa	66	76	3
Lamia	67	78	2	Mezium	45	54	7
Lampra, *Sup.*	74	3	2	*Miarus—see* Gymnætron.			
Lamprias	2	2	2	Miccotrogus	55	67	5
Lampyris	41	49	5, 6	Micralymna	13	15	5
Larinus	60	69	4	Micropeplus	25	30	5
Lasioderma	45	54	1	Microrhagus	38	46	6
Lathrobium	16	18	8	Miscodera	4	4	5
Latridius	28	33	8	Mniophila	70	81	6
Lebia	2	2	1	Molytes	61	71	1
Leiodes	22	27	6	Monochamus	66	77	4
Leiosoma	61	70	8	Mononychus	57	66	1
Leiophlæus	62	71	6	Monotoma	27	33	5
Leiopus	66	77	3	Mordella	47	56	1
Leistus	8	9	3	Mycetæa	26	31	6
Leptura	67	79	2	Mycetocharus	51	59	6
Lesteva	12	15	1	Mycetophagus	26	32	3
Licinus	7	8	5	Mycetoporus	18	22	2
Limnebius	32	39	1	Myeteras	53	61	7
Limnichus	29	36	2	*Mylæchus—see* Colon.			
Limonius	39	47	4	Myllæna	19		
Lissodema	27	33	4				
Lixus	60	69	5	Nanophyes	56	64	3
Lopha	9	11	4	Nebrin	8	9	5
Loricera	8	9	2	Necrobia	44	52	6
Lucanus	34	42	3	Necrodes	24	28	7

	Page.	Pl.	Fig.		Page.	Pl.	Fig.
Necrophorus	24	28	6	Pæderus	15	17	7
Necydalis	65	74	3	Pælobius	10	12	6
Nedyus	57	65	6	Panagræus	7	9	1
Nemoïcus	61	70	4	*Pangus—see* Selenophorus.			
Nemosoma.	28	34	4	*Panus—see* Thamnophilus.			
Nitidula	24	29	4	Paramecosoma	26	31	7
Nosodendron	30	36	4	Parnus	30	37	2
Notaphus	9	11	3	Paromalus	34	41	4
Notaris	59	68	7	Patrobus	4	4	7
Noterus	11	13	3	Pediacus	28	34	2
Nothus	51	60	2	Pedinus	48	57	2
Notiophilus	8	10	6	Pella	20	24	7
Notoxus	47	55	5	Pelophila	8	9	6
				Penichroa, *Sup.*	76	8	3
Obrium	65	75	5	Peryphus	9	11	5
Ocalea	20	24	4	Phædon	70	83	3
Ochina	45	54	2	Phalacrus	71	85	1
Ochthebius	32	38	4	Phaleria	50	59	1
Ocypus	17	20	5	Philhydrus	32	40	2
Ocys	9	11	6	Philochthus	9	11	7
Odocantha	2	1	6	Philonthus	16	19	8
Odontonyx—see Olisthopus.				*Philopedon—see* Cneorhinus.			
Œdemera	52	61	5	Phlæocharis	13	16	5
Oiceoptoma	24	28	8	Phloëobius, *Sup.*	75	6	1
Oligota	19	23	3	Phlæopara	20	24	1
Olisthopus	6	6	3	Phloiophilus	27	32	6
Omalium	13	15	4	Phloiotrya	51	60	5
Omaloplia	36	44	4	Phosphuga	24	29	2
Omaseus	4	3	8	Phylan	49	57	4
Omias	60	70	1	Phyllobius	61	70	3
Omophlus	51	59	8	Phyllopertha	36	45	1
Oncomera	53	61	6	Phytonomus	61	70	5
Onthophagus	35	43	5	Phytosus	14	17	2
Onthophilus	34	41	7	Pissodes	60	69	1
Oödes	7	8	2	Pityophagus	25	30	8
Oomorphus	30	36	6	Platycerus	34	42	1
Opatrum	49	57	5	Platydema	50	59	3
Opilus	43	52	3	Platyderus	4	4	2
Ophonus	3	3	3	Platynus	6	6	7
Orchesia	52	61	3	Platypus	64	73	8
Orchestes	58	67	1	Platyrhinus	54	62	5
Orectocheilus	12	13	8	Platysma	5	5	2
Orobitis	56	65	2	Platysoma	33	41	1
Orsodacna	68	80	5	Platystethus	14	17	1
Orthochætes	58	66	7	Plinthus	61	70	7
Orthoperus	22	27	2	Podabrus	42	50	4
Oryctes, *Sup.*	73	4	1	Pœcilus	4	3	7
Othius	16	19	2	Pogonocerus	66	77	2
Otiorhynchus	60	69	6	Pogonus	4	3	6
Oxyonus	35			Polistichus	2	1	3
Oxypoda	19	23	5	Polydrosus	62	72	3
Oxyporus	17	20	4	Polygraphus	63	73	6
Oxystoma	55	63	6	Polystoma	20	24	8
Oxystomus, *Sup.*	73	1	1	Poophagus	57	65	4
Oxytelus	14	16	8	Prasocuris	70	83	4
				Pria	25	30	1
Pachyrhinus	58	67	3	Prionus	64	74	1
Pachyta	67	79	4	Pristonychus	6	6	5

	Page.	Pl.	Fig.
Procas	61	70	5
Procrustes, *Sup.*	73	1	4
Proscarabæus—see Meloë.			
Prosternon	39	47	6
Proteinus	13	15	8
Psammodius	35	43	7
Psammœchus	68	80	6
Pselaphus	22	26	5
Pseudopsis	13	16	2
Pterostichus	4	4	8
Ptilinus	45	54	4
Ptinus	45	54	6
Ptomophagus	23	28	2
Ptosima, *Sup.*	74	3	6
Purpuricenus, *Sup.*	75	7	1
Pyrochroa	46	55	1
Quedius	16	20	1
Ragonycha	42	50	5
Raphirus	16	19	7
Remus—see Cafius.			
Rhagium	67	78	3
Rhamphus	55	63	5
Rhinobatus	60	69	3
Rhinocyllus	60	69	2
Rhinodes—see Thamnophilus.			
Rhinomacer	53	62	7
Rhinonchus	57	65	5
Rhinusa—see Gymnetron.			
Rhipidius, *Sup.*	75	5	4
Rhipiphorus	47	56	3
Rhisotrogus	36	44	5
Rhynchites	55	63	4
Rhyncholus	55	64	1
Rhytirhinus, *Sup.*	75	6	3
Rhyzobius	72	86	5
Rhyzopertha, *Sup.*	75	5	3
Rhyzophagus	27	33	3
Rhytidosoma	56	65	3
Rugilus	15	17	8
Salpingus	53	61	9
Saperda	66	77	1
Saprinus	34	41	5
Sarrotrium	50	58	8
Scaphidium	23	27	7
Scaphisoma	23	27	8
Scarites	2	2	3
Sciaphilus	63	72	6
Scirtes	41	49	2
Scolytus	64	73	3
Scraptia	52	61	1
Scydmænus	22	26	6
Scymnus	72	86	4
Selatosomus	40	48	3
Selenophorus	3	2	6
Serica	36	44	3
Sericoderus	22	27	1
Sericosomus	39	47	2
Siagonium	15	18	5
Sibinia—see Sibynes.			
Sibynes	58	67	4
Silis	42	50	2
Silpha	24	29	1
Silvanus	28	34	1
Simplocaria	30	36	7
Sinodendron	34	42	4
Sitaris	48	56	4
Sitona	63	72	4
Sogines, *Sup.*	73	2	1
Sphæriestes	53	61	8
Sphæridium	33	40	5
Sphæroderma	70	83	2
Sphærosoma	72	86	3
Sphærites	23	28	5
Sphærula—see Nanophyes.			
Sphindus	44	53	1
Spercheus	31	37	6
Sphodrus	6	6	6
Spondylis	64	74	2
Staphylinus	17	20	8
Stene	49	58	4
Stenolophus	3	3	4
Stenus	14	17	5
Steropus	4	4	3
Stomis	4	4	6
Strangalia	67	79	1
Strongylus	24	29	6
Strophosomus	63	72	7
Sunius	15	18	2
Sybaris	48	56	5
Synaptus	40	48	6
Syncalypta	29	36	3
Synchita	27	33	1
Synuchus—see Taphria.			
Syntomium	13	15	7
Tachinus	18	21	4
Tachyerges	58	66	7
Tachyporus	18	21	7
Tachypus	9	11	2
Tachys	9	12	1
Tachyusa	20	24	2
Tænosoma	13	16	4
Tanymecus	63	72	5
Tanysphyrus	62	71	3
Taphria	6	6	4
Tarus—see Cymindis.			
Tasgius	17	20	4
Telephorus	42	50	3
Tenebrio	49	58	3
Teredus	28	34	6

	Page.	Pl.	Fig.		Page.	Pl.	Fig.
Teretrius . . .	34	41	6	Triphyllus . . .	27	32	5
Tetropes, *Sup.* . .	76	8	1	Tritoma . . .	71	85	3
Tetratoma . . .	26	32	2	Trogophlæus . .	14	16	6
Tetrops	66	76	1	Trogosita . . .	28	34	3
Thamnophilus . .	60	68	8	Tropideres . . .	54	62	4
Thauasimus . . .	43	52	4	Trox	36	44	2
Throscus . . .	28	35	1	Trypodendron . .	64	73	5
Thyamus . . .	69	82	2	Tychius . . .	59	67	6
Thymalus . . .	24	29	3	Tychus . .	21	26	1
Tilloïda . . .	43	52	2	Typhæa . . .	26	31	5
Tillus . . .	43	52	1	Typhæus . . .	35	43	2
Timarcha . . .	70	83	6				
Tiresias . . .	29	35	3	Uleiota . . .	64	73	10
Tomicus . . .	64	73	7	Uloma	49	58	2
Toxotus . . .	67	78	4				
Trachyphlæus . .	61	70	2	Valgus, *Sup.* .	74	4	3
Trachys . . .	38	46	4	Velleius . . .	18	21	3
Trechus . . .	7	7	7				
Tribolium, *Sup.* .	73	2	5	Xiletinus . . .	45	54	3
Trichius . . .	37	45	5	Xylophilus . .	47	55	3
Trichopteryx . .	25	30	4	Xylotrogus . .	28	34	7
Trimium . . .	21	25	7				
Trimorphus . .	7	8	7	Zabrus . . .	5	6	1
Triuodes . . .	29	36	1	Zeugophora . .	68	80	4
Triplax . . .	71	85	4	Zyras	20	24	6

www.ingramcontent.com/pod-product-compliance
Lightning Source LLC
Chambersburg PA
CBHW021515210326

41599CB00012B/1258